# Layout Command Control 2025

## *Basic and Advanced Principles of LCC for Model Railroads*

Dana W. Zimmerli, PhD

ZSysDesigns, 2025

Dana W. Zimmerli
dana@zsysdesigns.com

First Edition: March 2025

The author is grateful for permission to use
the following previously copyrighted material and trademarks:
Dick Bronson, RR-Cirkits Trademarks and excerpts from the
CDI as displayed in the Configuration Tools.
Bob Jacobsen of JMRI for his advice on the treatment of
JMRI screen images

ISBN: 978-0-9888259-3-2
Printed in the United States of America

# Contents

# List of Figures

## List of Tables



# Dedication

This book would not have been possible without the support and tolerance of my wife. She encouraged me in the days when it seemed impossible, and Edited every chapter so that at least the English is understandable.

I also appreciate the support as a photographers assistant that Mike Wissler provided.

# Introduction

This book was inspired by my original book, Introduction to Layout Command Control. After giving clinics on LCC for two years in the Pacific Southwest Region of the NMRA, I realized that the beginner would need something to refer to and i could not reach everyone in clinics. Constant messages on the LCC group made it more clear. I decided that I could write this down so that all the hobbyists that wanted to implement LCC would have a starting point.

Since the self-publication of that book, I have heard many persons comment on the aid that book provided in using LCC. After 5 years, LCC has changed to the point that an update is needed. In addition, some are looking for some more advanced information. This book is the result.

I have spent my entire career through to retirement in aerospace development. As a System Engineer, Technical Director, and Program Manager, I participated in several proposals for military contracts. I decided that, if I could translate engineer-speak for the military, I should be able to explain LCC to Model Railroaders.

I use the RR-CirKits™ LCC Products in most of this book. That is because RR-CirKits is the primary manufacturer with a full range of products. I have included some manufacturers that provide variants of these Nodes with more, less, and different I/Os. The basic principles for LCC should not be very much different, regardless of the manufacturer. In Chapter 3, a minimal LCC Network is described. If the reader wants to follow the steps (highly recommended) then the best way to do that is to obtain a Starter Kit, a Tower-LCC and a Test I/O Board from RR-CirKits. That will allow the reader to set up an LCC network with minimal effort.

While LCC is intended to operate without a computer (as in a desktop or laptop), the initial setup (called configuration) of the LCC network is performed on a computer. The examples in this book are screen captures from the JMRI configuration tool, running on a Microsoft Windows 11operating system. JMRI is easily obtained, is free and runs on most operating systems. I have personally run JMRI on a Linux system, as well as the Windows system. I have limited the examples in this book to JMRI even though there are other programs available. That is because JMRI can run on almost any system.

JMRI is used in this book for the features of JMRI related to LCC, but I do not discuss the many other features of JMRI. For example, JMRI has a very complete signaling implementation which can be used to complement LCC. I do not discuss this case, rather, I implement signaling with LCC boards. Use of JMRI for this purpose would require a permanent computer connection.

What I decided was needed was a book combining the basic introductory material with more advanced discussions. The introduction to LCC Principles is in Chapter 1. This chapter sets the terms and ideas in place for later chapters.

Chapter 1 is written a level above the NMRA Standards. In other words, I try not to discuss details to the bit level or even details about the messages. A conceptual knowledge is the goal of the first chapter. I also intend to provide a practical guide to implementing the LCC system. Because of that, I will not discuss the basic standards for LCC. This is a crawl before you walk book, but walking is the goal. Many features of LCC are not covered, primarily because implementing LCC for a layout does not require those details. If someone needs the details, they can access the Standards and Technical Notes on the NMRA website.

The second chapter is a slight diversion from the strict discussion of LCC. To effectively use LCC, some knowledge about the devices connected to the LCC devices is needed. Even though my PhD is a double major in Computer Science and Electrical Engineering, I tried to keep the details to a low enough level that most hobbyists would understand. If you are an Electronics buff, you can probably scan this chapter. If not, all I ask is that you know how to add, subtract, multiply and divide — with a calculator, of course.

As mentioned above, Chapter 3 assembles a minimal LCC network to learn some of the basics. Using the recommended devices, no electronics assembly (soldering) will be required. All you need to do for this LCC network is plug in the cables. Chapter 3 also presents some advanced features of the basic I/O Nodes from RR-CirKits. These features were not available in boards when the first book was published.

Chapters 4 – 6 step through the design and implementation of an LCC network based on a real example. Designing the network, documenting the parts, and configuring the devices is covered with many examples.

Many modelers want to implement signaling on their layout, if only for the "eye candy" reason. Chapter 7 is my attempt at demonstrating how to implement signaling using LCC. I definitely am not an expert in signaling, and I am certain that my attempt at this is only a very simple start, but I hope to show how the signaling can be implemented. If anyone thinks that something is extremely wrong, let me know. To get into signaling, I have started from the basics of logic operations, since these operations are critical to implementing signals. The example for signaling builds on the demonstration layout in the earlier chapters.

More signaling functions are now available. Chapter 8 explores the use of these signaling functions, with an example derived from a real layout. These new features extend the signaling to more signals, without using so many Nodes. Other signaling methods are being developed, some which interface with LCC. These are not included. I intend this to be only LCC.

Chapters 9 and 10 explore the use of LCC Fast Clocks and how to use them in realistic lighting scenarios. Combining a time with the lights on a layout will attract attention and make the layout more realistic.

Chapter 11 is a first attempt to explore the use of LCC in train control. New products implementing train control are just coming out, so this is a very dynamic time for this area.

An interesting thing about life and model railroading is that people often trade time for money. The old saying is "Time is money," and that is true in model railroading. in our hobby, it applies to the question "what would you rather do, X or Y?" In particular, for LCC there are often methods of accomplishing the same thing by purchasing some more expensive parts. A common example is connecting to Tortoise turnout machines. If you don't like to solder on leads while looking up under the layout, you can buy a connector that has screw terminals — for a higher price than regular connectors. Many examples are available. In most of this book, I have opted to show the basic methods which are often more work. In Chapter 12, I will explore some of these examples in the chapter I call "Present and Future Options."

In the 5 years since I first wrote the book, LCC has grown significantly. Several new manufacturers are making LCC product. Small to large layouts are using LCC for turnout control and realistic signaling. New uses for LCC are being developed and the capabilities of LCC nodes are expanding. At the recent NMRA convention the LCC oriented clinics were standing room only. These are all indicators that LCC is here to stay and has a role in model railroading.

Finally, I do not claim to be perfect. I have noted that baseball hitters are considered great if they can hit close to 40% of their at-bats. I hope to be better than 50%, since that means I am right more than I am wrong. Nonetheless, I am certain there are errors in this book. So, please excuse the human. If the errors are significant (typos that do not change the meaning of sentences are not significant) send me an email. It has been an exciting time so far and I expect it to continue. New products will be popping up at any time. I heard that "Change is the essence of life." Let's enjoy life and all its changes.

Dana W. Zimmerli, February 2025

dana@zsysdesigns.co

# About the cover:

The cover photos were taken at the Belmont Shore Model Railroad club in San Pedro, CA. The front cover depicts the Interlocking section at the end of the main yard, where two mainline tracks, two yard leads, and the spur to the harbor all come together. This is a challenge for the future implementation.

The back cover is a shot down the mainline in the town of Echo. The center passing track is the source for the examples in Chapter 8.

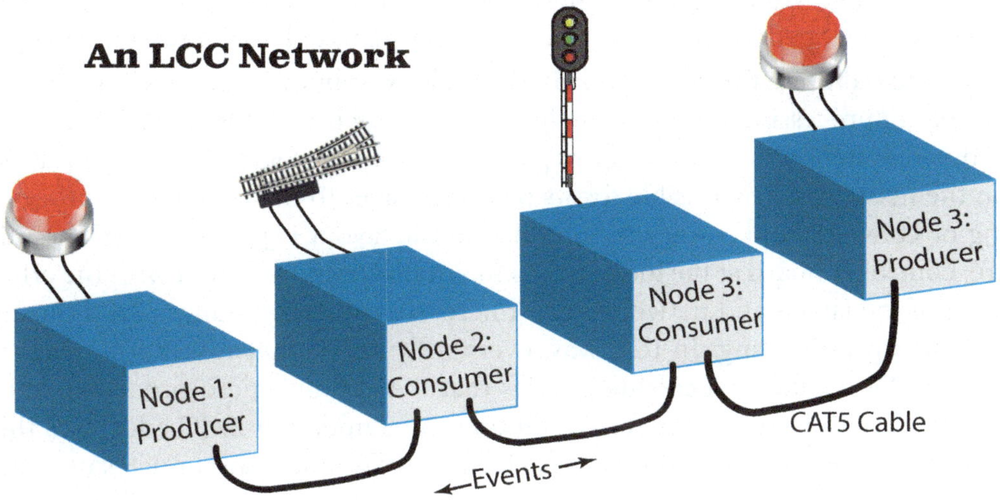


# Chapter 1 – What is LCC?

Layout Command Control (LCC) has been called many things: An information super-highway for model railroad layouts; A common language for layout elements to talk to each other; An Open NMRA Control bus standard for hobbyists and manufacturers; and several other things we won't mention. In this chapter I will attempt to describe what LCC is from a non-technical viewpoint — well maybe a little technical, but only LCC installation and all the associated layout parts. To do that I need to explain details, when it is necessary. The goal is to understand enough about LCC to plan the network. To do that, some basic principles on which LCC was created follow in this chapter.

## What is LCC?

LCC is a distributed system that uses modern technology to control the layout. This statement has several pieces that need further e4xplanation:

**Modern Technology:** First, the technology that supports LCC is in everyday life —every automobile sold today has a bus to control all the functions. This bus is the Controller Area Network (CAN) bus. It connects the engine to the dashboard, the dashboard to the lights, the radio to the GPS, and every other device in the car. Each of these devices is controlled locally by a small micro-controller that listens to the bus and reacts to commands sent to it. That micro-controller is a small integrated circuit (often referred to as a "chip") like the ones found in model railroad locomotive decoders. Both of these technologies are very mature and have been in use for many years. LCC does not require a specific technology to be used. In fact, other types of network are being developed for LCC. Ethernet, Wi-Fi, and other networks may be available in the future. When they are developed for LCC, they will probably be based on mature technology.

**Distributed Processing:** This technology supports the LCC system by providing processing power at the locations where it is needed. LCC is a a distributed system which does not depend on a master computer. Each location on the network is responsible for its own interfaces to the layout. Any processing or conversion necessary to interface to the layout is performed in the micro-controller board at the location. To maintain communication with the other parts of the layout and network, each micro-controller announces itself during its start-up processing. In that way, everybody on the network knows who is present. Distributed processing implies that the various micro-processors may not be physically accessible for configuration or updating. LCC addresses this situation by providing methods to perform configuration and updates over the network.

**Network Connections:** To create a distributed system there must be a network to tie it together. As noted above, the first LCC was connected with a CAN. Although other forms of network are possible (and planned) the CAN provides several important features:

1. It has a single cable (2 or 3 wires)to connect;
2. It is inherently a broadcast network, which means that every device on the network sees every message;
3. It has built-in error handling to provide reliable communications;
4. Some "housekeeping" messages are already defined.
5. It ensures that messages are transmitted even if the network is congested.

These features allow the LCC system to be connected with minimum wiring between local areas. This reduces the number and length of wires , which improves reliability and makes maintenance easier. This solution is very similar to what the prototype railroads did in the early 20th century. So LCC can be thought of as just another case of modeling the prototype.

The definition of LCC is contained in the NMRA standards. This set of standards is a technical description of the LCC requirements. Details of the electrical and network requirements are needed to accomplish the features described above. The LCC standards are similar to the DCC™ standards in that few people have to read and understand the standards. To ensure compatibility between different manufacturers, the standards are necessary, but like DCC, most hobbyists will never need the standards. The information in this book will cover the functions of the requirements in the standards documents, attempting to stay at a level that hobbyists understand. The important point to know is that the standards only define the network interaction requirements, not the interface with the layout. More on that in later sections.

Before we finish describing what LCC is, we need to state what LCC is not. LCC has been described as controlling everything that is not on the track. It is not intended to replace DCC. Control of the locomotive and cars on the track will continue to be provided by either DCC or DC on the track, In some instances LCC may interact with DCC to provide additional controls (for example, an interface to a DCC control station could provide translation for many types of throttle.) Similarly, LCC does not depend on DCC. LCC and DCC operate independently of each other, and normally are not even aware of each other.

# A Short LCC History

As DCC became popular more features were added unrelated to the control of locomotives. Accessory controls were added for functions such as turnout control, signaling and scenery lighting control. These were transmitted on the DCC bus (or on a parallel DCC-like bus beside the track) to interact with trackside devices. As larger layouts used more features, the impact on locomotive control was noticed. In addition, locomotive sound systems became popular, and additional messages were required to support these functions. The impact of these messages became noticeable as delays in locomotive control. In some cases, messages on the DCC lines were lost and locomotives were lost. The problem was identified as too much traffic on the DCC bus. NMRA recognized this issue and began to investigate alternate networks for layout control.

The current LCC standard was developed by several hobbyists from several different countries. This group is the Open Layout Control Bus (OpenLCB) group. They proposed their solution to the NMRA and it was chosen as the preferred network for layout control. The proposed standards were approved by the NMRA in 2016. When the NMRA adopted the standards, they renamed the system Layout Command Control (LCC.) At this time, both LCC and OpenLCB exist in the documentation. These standards covered the use of the CAN bus for LCC. Additional standards are anticipated to cover other network implementations and to add functions to the messages that are currently defined.

Commercially available LCC devices are being developed, with the primary source being RR-CirKits™. RR-CirKits has several boards available to implement LCC on a layout. Some of these are used in this book as examples. Other manufacturers used in this book are Logic Rail Technologies™ and Train Control Systems™. In addition, LCC devices have been developed on some micro-controllers for development purposes. Eventually, a DIY LCC device may become available for the hobbyist interested in developing special functions.

As we will see later, LCC is designed to contain its own configuration data. Configuration data provides the device with information about the interfaces to the layout and the messages that it needs to process. This configuration data is set using a Configuration Tool (CT). Two CTs are currently available, one based in the populate JMRI™ software, and one supplied by Deep Woods Software. These tools are constantly being enhanced to improve the ease of configuring LCC.

# Potential Applications of LCC

The applications for LCC are still evolving as LCC evolves. Several layouts are currently applying LCC for controls and signaling. Additional applications recent;y implemented include fast clock implementation and traction controls. Many other possibilities can be imagined. Applications are only limited by your imagination.

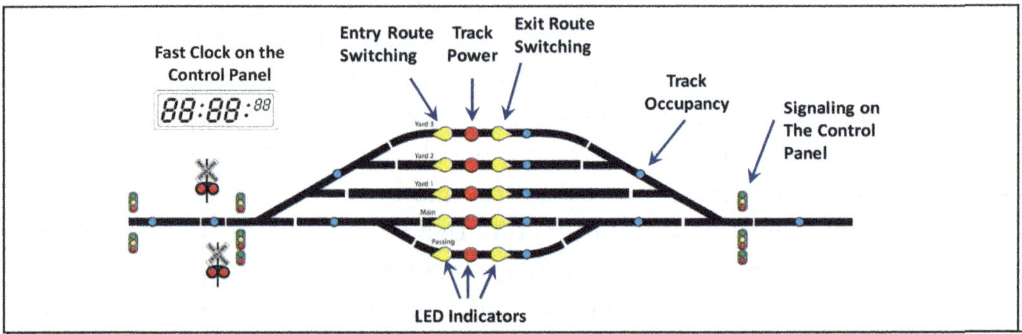


**Figure 1.1 Control Panel Application Shows Multiple Functions.**

Figure 1.1shows several features implemented with LCC. Each one may be connected to either a local or a distant interface. For example, the buttons that switch the turn-outs for routes may control a turnout many feet away. Indicators for occupancy will be distant from the tracks that are occupied. In addition, this panel is anticipated to control a yard which extends below another part of the layout, so engineers operate it from both sides of a peninsula. To prevent constantly walking around this peninsula, a duplicate panel will be added. In LCC this simply means copying the panel and LCC devices. No long wires will be needed to duplicate the panel.

Looking at the pieces in a bit more detail, we see the following:

1. Each ladder track has two buttons for route switching — one button for each direction creating an NX (eNtry-eXit)control.
2. Another button on the ladder tracks controls the track power — A relay on the track feeders allows locomotives to be parked without fear of run-aways;
3. Each button has two functions —to indicate a condition and to cause a change;

4. Signaling is shown on the control panel, which makes it convenient to choose tracks and keep running;
5. A clock face on the control panel can show the fast time clock — each control panel indicates the current time without looking at some other location, This feature could be extended to several wall clocks, different wall clocks on different time zones, and even a transmitted clock time to implement individual watches for engineers.

**Figure 1.2 Structure Lighting adds Realism to the Layout**

Additional uses for a fast clock might be to control the lighting in scenery structures. Internal lights and street lights in towns could be turned on and off at times according to the clock currently in use. This added piece of realism could then be extended to control the layout lighting to match the time on the clock in use. As I said before, the applications are only limited by your imagination.

## Basic Terms and Concepts of LCC

Several terms are tossed around when talking about LCC. The one term that is used constantly in LCC is "event." Other related terms are "Producer," "Consumer" and "Nodes." These can be confusing to the non-technical beginner. It is important to learn some of these terms to understand LCC Let us start with the most common term, the Event:

### An Event reports a change in the state of something.

Almost everything on a layout can be described by its current state. The state of something is simply the condition it is in. For example, a turnout can be described as "normal" or "diverging." These two words tell us what the state is. Other examples include a push button, which can be either "open" or "closed", a track which is either "empty" or "occupied" and an indicator light which is either "on" or "off. Some states may have a value, such as a light which may be dimmed, but then the state is defined by its value.

Unfortunately, a network is not designed to transmit states continuously. That would require either infinite band width or many wires, maybe one per state. If that sounds familiar, that is because that is how we have wired layouts for years — connect one wire to each state that needs to be controlled or indicated and then connect that wire to the other device. This method leads to long wires all over the underside of the layout.

An Event can be defined as a report of a state change. When a push button is pushed, an Event occurs; when that push button is released, an Event occurs. In LCC we may not care about all of the Events that can happen, but we could. Events we care about are inserted into messages on the network and sent out to anyone that wants to know about them. Some details about Events will be covered later, but that is the source of Events. The reason we use Events in LCC is so that the network does not have to constantly monitor all the states on a layout. Events take less time on the network, since they only appear when a state changes.

### Producers create Events; Consumers act on Events

A simple way to think about the interactions on an LCC system is the Producer/Consumer model. This model defines the creation and use of Events and provides a common reference for LCC. Figure 1.3 shows the variation in Producers and Consumers allowed in the model. As we will see in later chapters, the devices connected on the Producer or the Consumer can vary greatly in their electrical needs. The LCC standards do not directly address the inputs and outputs, only the network.

Associating the external I/O with a particular producer or consumer and the EventID is performed by the Configuration Tool (CT). Configuration is the process of connecting the I/O on a board with Events and defining what action is taken. Since the action is to change the state of some layout feature, the Consumers have to remember the state that was last set. Similarly, the Producers often have to remember what Event was last sent. One of the network messages provides a way for the Consumers to ask what the last Event was, so that the consumer can "catch-up" during start-up.

As shown in Figure 1.3, multiple Consumers can be connected to an Event. Multiple Consumers use Events like switch closure to change a turnout. Consumers for this Event might be the turnout motor, the LED that indicates

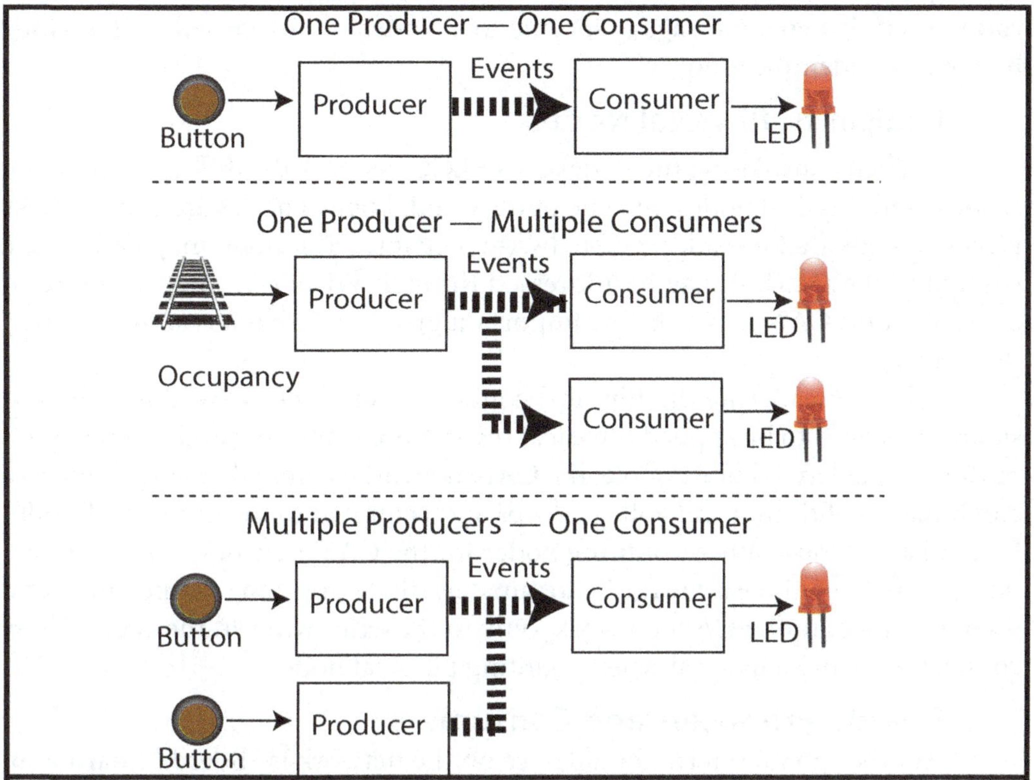


**Figure 1.3 The Producer-Consumer Model is Flexible.**

the turnout position and the signals that control the track segment.

Similarly, multiple Producers can appear, although it is less common. Duplicate control panels is an application where multiple Producers will be used. Multiple Producers can be dangerous, because the system can get the same Event or conflicting Events from different locations. Deciding which to use is not clear, Various ways to "lock-out" Events may be necessary.

## LCC Connects Nodes on the Network

The basic unit on the LCC network is a Node. The definition of a Node in the standards is somewhat ambiguous. In the standard for the CAN Hardware, a Node is a physical connection to the network. This is understandable and is something that can be touched and seen. In the hardware standard several physical requirements are stated. Things like voltage, current, connector type and pin functions are defined. These physical characteristics are necessary to design an LCC system. The ordinary hobbyist who is assembling an LCC system will be very aware of these requirements.

In all of the other standards for LCC, a Node is defined as an addressable software process. The definition explains that multiple Nodes may exist in a computer, but every board connected to LCC must have at least one Node. All messages on the LCC network are created by a Node and has the Node identification on it. When a message is specific to a Node, it is addressed to that Node by the Node identification.

## Logical vs. Physical Nodes

When considering the nodes on an LCC network the difference between Logical and Physical nodes must be understood. Logical nodes are simply those pieces of code that are addressable by the network. The node may be inside a computer like JMRI, it may be addressed through Wi-Fi or it may be a microcomputer on a CAN network. The important property is that it is addressable by the network.

When considering the Physical connections to a network, different considerations need to be applied. Physical restrictions on timing and voltage levels must be considered. For example, the CAN network is limited in length due to a combination of timing and voltage drops and is limited to the number of nodes for similar reasons. When counting nodes for the CAN network, every connection seems to be the easy way., Unfortunately, there are some connections that do not connect to the CAN network, only to the extra wires in the cable. These connections can be ignored when counting physical nodes.

## Boards, Producers and Consumers

Another popular term for a device on the network is "Board." Board is an indeterminate term, since it usually contains specific hardware to connect to the layout. In Figure 1.4, an example is shown. This Board has both Consumers and

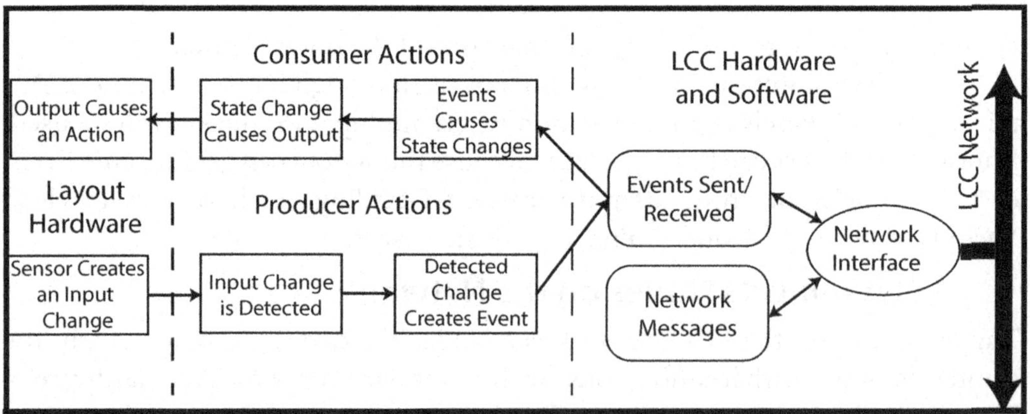


**Figure 1.4 Boards may contain multiple Producers and Consumers**

Producers on the same Board. On the right side, the LCC Network connects into the Board through a network interface. This network interface may be any hardware — the current interface defined is the CAN. The primary definitions in the standards are concerned with the physical interface (connectors, cables and pin assignment on the connectors.) These will be discussed in a later chapter.

Moving in from the right, the next blocks are the required LCC software. Message handling, Event structure, start-up, and configuration are included in this section and are defined by the LCC standards. This is the extent of the LCC requirements on a Board.

The additional blocks in Figure 1.4 are concerned with the interface to the layout. For a Consumer, LCC delivers an Event to the Consumer and the Consumer has to decide what to do with that Event. Typically, the Consumer has been configured to interpret that specific Event into an Output. For example, the receipt may mean to turn on an LED on pin 3. The Consumer would then know (from configuration) that pin 3 must be "set" to turn on that LED, so this piece of software will set pin 3 in response to receiving this Event. (Event Identification will be covered shortly.) The Producer operates in reverse. When an input changes. There is a change detected on a pin. The Producer software recognizes that change and associates an Event with the change. The specific identification of the Event is stored in memory during configuration. Using the stored information, the Producer creates the correct Event and sends it to the LCC software for transmission to the network.

Since a Board can contain multiple Producers and Consumers, and since the Board is designed specifically to interface with specific layout hardware, it is difficult to define a Board. Most hobbyists will think of a Board as a physical thing which contains electronic devices connected on a printed circuit board. However, it is convenient to refer to this Board as a Node. That is incorrect. In the NMRA Standards, a Node is either the software in the rounded boxes of Figure 1.4, or the hardware labeled "Network Interface." To further complicate the names, the configuration of the Board is defined by functions required in the LCC standards.

To complete this discussion of Nodes, let me simplify the rest of this book by incorrectly using the term Node for all of the hardware and processing on an LCC Board. If the distinction becomes necessary, I will refer to the Hardware Node or the Software Node. When the term "Node" is used, that will refer to the entire "Board" installed on the layout.

## NMRA Standards and Conformance

One of the main reasons for NMRA Standards is the Conformance program which ensures that a product meets the NMRA Standards. This program is maintained by a process of testing to check that the Standards are met. The result of the testing is a conformance certificate and the ability to use the Conformance symbol on the product. Many DCC products have gone through this process.

Considering the difficulty of defining what constitutes a "Board" and the fact that no one will use only an LCC node, it is impossible to test a Board for NMRA Conformance. At best, some part of the Board can be tested to show that it can be configured and it sends and receives Events. Some protocols can be tested because they have specific responses to defined messages. More than this will be very complicated and require a significant investment in equipment to cover all I/O.

Until recent additions, LCC did not interact with external I/O, so measuring an action was dependent on the manufacturer's documentation. Most of the Producer/Consumer events either start with an external device or change an external device state. These interactions are not normally defined by the LCC Standards.

As a side issue, it is very difficult, if not impossible to understand the Standards without reading the Technical Notes (TNs) at the same time. The TNs expand on the requirements in the Standards for understanding, but they are not requirements.

Since the unit produced is the board, and since LCC only defines the network required, Conformance testing can only verify the interface to the network, not the interface to the external device. If such a limited type of Conformance testing is established, only part of a board is in conformance with the Standards. If modelers purchase a board which has only partially verified, won't they assume incorrectly, that the NMRA stands behind the working of the entire board?

## Globally Unique — Node and EventIDs

To ensure the ability to add Nodes without conflict between Nodes, Nodes are given a "Globally Unique Identifier." This allows a form of "Plug-and-Play" because when a Node is plugged into the network, it cannot have an ID which is already on the network. This is implemented by assigning each Node a 48-bit Identification. Using 48 bits creates over 281 trillion ($2.81 \times 10^{14}$) Identifiers. Although this is such a big number it is hard to conceive, the use of that set is reduced by assigning portions of it to specific uses. To ease the reading and writing of these numbers, the identifiers are written in 2 byte hexadecimal (16 bits) chunks separated by a decimal point (period.) This format will be seen anytime a Node ID, EventID or Event is displayed.

> **Hexadecimal**
>
> The easy way to get hexadecimal numbers is on your computer. Most modern PCs have an app for a calculator. One of the optional calculator types is a "Programming" calculator. Opening this calculator will reveal two or three fields label "DEC' and "HEX". If you enter a value in either one, the calculator converts that value to the other type and displays it.

Node IDs are allocated in groups identified by the first byte. Several of these groups remain unallocated or reserved for special purposes. The details are in the standard document S-9.7.0.3 LCC Unique Identifiers. The most used ranges are:

- 01.nn.nn.nn.nn.nn which defines "Well known Global Identifiers" — these are used for some of the newer protocols to provide global events. When used the events are then restricted to very specific meaning.
- 02.nn.nn.nn.nn and 09.nn.nn.nn.nn which identify the Manufacturer ranges — These IDs are for the LCC manufacturers. A section of them are assigned to known DCC manufacturers by their NMRA number. The last 3 bytes of these numbers are left for the manufacturer to assign, allowing a manufacturer to assign 16 million node IDs before needing another range.
- 03.00.nn.nn.nn.nn is assigned to NMRA members for personal use. The third, fourth and fifth bytes contain the member's NMRA number, and the sixth byte is left for the member to assign, giving each member 256 IDs.
- 05.01.nn.nn.nn.nn.nn are assigned automatically by request. The last byte is assigned sequentially by the person owning the range. This provides 256 IDs for each number requested.

As described above, the assignment of IDs is easy and provides a large number of globally unique identifiers for Nodes. Globally unique identifiers is one of the basic concepts of LCC.

| **Node Identification** | | | | | | |
|---|---|---|---|---|---|---|
| Byte 1 | Byte 2 | Byte 3 | Byte 4 | Byte 5 | Byte 6 | **48 bit Node ID** |
| 02 | 01 | NMRA DCC | Self Assigned | | | **NMRA DCC Maanufacturers** |
| 03 | 00 | NMRA Member Number | | | Self Assigned | **NMRA Member Number** |
| 05 | 01 | Assigned by OpenLCB Request | | | Self Assigned | **Assigned by OpenLCB** |

| Unique Node Identification | Byte 7 | Byte 8 |
|---|---|---|
| **Unique Event Identification** | | |

**Figure 1.5 Summary of Unique Identifier Creation**

From the unique Node ID, unique EventIDs are created. Using the Node ID as the first 6 bytes, an EventID with an 8 byte length can be created. Each

Node ID can create over 65 thousand EventIDs. This design is summarized in Figure 1.5 which shows the progression from unique Node Identifiers to unique Event Identifiers.

While this connection is convenient, there is no requirement that EventIDs be dependent on NodeIDs. Using the NodeID as the first 48 bits of the EventID makes it easy to ensure that the EventID is globally unique. Since the NodeID is unique and assigned to the specific node, only the last 16 bits need to be assigned for a specific purpose. There is no check to prevent arbitrary EventIDs. A user is free to create any EventID, using any rule they desire. Using the unique NodeIDs listed above is a simple way to ensure globally unique EventIDs.

Typically, Producers create EventIDs, but allocating EventIDs to Consumers when manufacturing a Node ensures unique Events in the Node. This does not require that two Nodes cannot generate the same EventID, rather it requires that the same EventID is used for the same meaning any where it is generated. In my previous example of duplicate control panels, the same EventID would be generated by both panels, but would have the exact same meaning. This is acceptable.

### Well Known Events

The first set of "Node IDs" defined above are not actually associated with a node, but rather defined by their common meaning. All EventIDs that start with 01 in the first byte are reserved as "Well Known Events." The LCC Glossary defines these events as "globally defined to have specific meanings." The Standard that reserves these events does not define the specific meanings. Many of the Well Known Events are defined in protocol standards like the Broadcast Time Protocol. These standards use the Well Known Event and add the last two bytes (16 bits) to specify specific actions. A later chapter will discuss this protocol.

Other Well Known Events are not specifically defined anywhere. Some discussion of the intent is contained in the Tech Notes, but the specific implementation is left unstated

## LCC Network Messages

The LCC network is defined by the messages that it uses to control communications and to move data between Nodes. The developer of LCC hardware needs to know the precise format and content of these message. This information is contained in the standards. A hobbyist installing an LCC system is not likely to need the precise information in the standards, but a general knowledge of the messages and their use adds to the understanding of what LCC can do. Messages in LCC have specific formats which are common to every network implementation. Some implementations may require slight modifications to the messages due to limitations of the network. All implementations include basic message data, including a message type, a source ID and data specific to the message type. The messages that exist on the LCC network are grouped into categories below.

## Protocols — a Logical Grouping of Messages

Messages are defined in groups called Protocols. A protocol is a set of messages that accomplish some specific action, such as moving a specific kind of data between Nodes. In the standards, several Protocols are defined. These include the Event Exchange Protocol, the Configuration Description Protocol, and the Memory Configuration Protocol. These protocols may be supported by lower level protocols such as the Datagram Protocol. Additional protocols are defined in the standards and a number of protocols listed are yet to be defined. Definition of these protocols will add messages to the current list.

To define a "Simple Node," a subset of the entire Protocol list has been identified. This Simple Protocol list includes the important protocols to create a Node that performs the basic functions of LCC.

## Network Support Messages

These messages are needed to establish and maintain the network. They include Initialization and Node Identification messages.

- Initialization messages announce the presence of the Node to the network. The Node sends this message to the network when it has finished its initialization and is ready to receive messages. Included in this message is information defining whether it needs the full protocol set or just the Simple Protocol.
- Node Identification Messages — These messages provide a method to determine the Node IDs of the Nodes connected to the network. Either a single Node can be verified, or the message can request that all Nodes reply.
- Protocol Supported Requests — The request message is sent to all Nodes requesting reply identifying the Protocols supported. Each Node attached to the network responds with a message defining the Protocols supported. In some advanced applications it may be necessary to only send some messages to the Nodes that support those messages.

## Event Transport Messages

Since the primary purpose of LCC is to pass Events on the network, Event Transport messages are very common. Events are passed in the Producer/Consumer Event Report Message.

- The Producer/Consumer Event Report(PCER) — This message is used to pass the EventID to the Nodes on the network. Ordinarily the destination is all Nodes, although it can be directed to a specific Node. This is probably the most frequent message on an LCC network.

- Producer/Consumer Identify Request Messages — This set of messages ask the Nodes on the network for the specific Events that they Produce or Consume. The request can be either to a specific Node or to all Nodes. When all Nodes are requested, the traffic on the network may be very high. The Event part of this message can be either a specific Event or an Event Range. If the Event is fully identified, the responder must also respond with the current validity of the Event. If the Event is consistent with the current state, then it is considered valid. It can also be Invalid or Unknown.
- Producer/Consumer Identified Messages — This set of messages announce to the network that specific Producers, Consumers or ranges of Producers or Consumers are created or used by a node. These messages can be in response to an Identify Producer/Consumer message but are also required at start-up and when changes occur

### Datagram Messages

A datagram is a free-form message which is passed with only a header and length information. This is a way to pass large amounts of data on the network. It is very important for the Configuration task described in the next section. If the data is longer than one message, the data is split into multiple messages which are numbered so they can be re-assembled after receipt.

## Configuration and Memory Access

Earlier I noted that LCC requires data embedded in the Node to aid the configuration of a Node. I define configuration as the process of associating Events with I/O and specifying the form of the I/O required. In the next chapter I will discuss the different electrical forms that you might encounter and how to hook them up. Contrary to some opinions, configuration has nothing to do with programming. Configuration is closer to the assignment of values to DCC decoders than writing programming code. In LCC, the configuration process is performed using a Configuration Tool (CT) which reads data from the node and then formats it into a human readable display for input of options. Description of CTs will be left for a later chapter. This section describes the functions required by LCC to allow configuration.

### Configuration Description Information

As required in the standards, an LCC node must have some data embedded to describe it and its configuration. These data comprise the Configuration Description Information (CDI.) The data required in the Node includes the manufacturer, the version number of the firmware, and the version number of the hardware. These items are static and will not change. Additional data which may be static would be the number of I/Os, and the types of I/O available.

Variables in the configuration data will include the EventIDs associated with a specific I/O pin, what type of I/O it is, and how to interpret the action. Producers will associate the change of an I/O with an EventID to send; Consumers will associate a received EventID with an output on a pin. Other data allowed in the CDI may include descriptive information to name the Node and to name specific pin functions. To support these variables, the variable types defined by the standard are EventIDs, integers, and strings. This determines the type of data the user can enter.

It is up to the designer of the specific CT to format this data into a human format for display. LCC Nodes store this information in a coded format using the Extensible Mark-up Language (XML.) XML includes addresses to specify where in memory to find/store the data, as well as descriptive data to display for the user. The messages that read and write the CDI are contained in the Memory Configuration standard.

### Memory Access

Nodes typically have two or three different memory spaces to perform their tasks. The main processing of a Node will require memory for the program, memory for the program variables, and memory for the CDI. Even the CDI may exist in different types of memory, using one memory space for the static parts and another for the variables. Access to the CDI depends on the Memory Configuration protocol. Memory Configuration defines three memory address spaces which are allocated by the Node to contain the CDI definition, the CDI variables, and everlasting (as defined by the designer.) Read and Write messages are constructed using Datagrams to carry the data on the network. This is the primary use of Datagrams in this revision of the standards.

## Firmware Upgrade

There are always errors in firmware and there are always things that can be improved. The distributed design of the LCC network results in many µCs located all over the layout. Furthermore, the nodes which contain the µCs are usually difficult to access and would require disconnecting many cables. Fortunately, the LCC network is connected to all of the nodes, so a special protocol was developed to load firmware updates through the network

The Firmware Upgrade method relies on the Configuration Tool to read a file from a computer and transfer it to the designated node. Leaving out the details, typically the CT starts the upgrade by notifying the node that a firmware upgrade is to be performed, then transfers the new firmware. The CT then transfers the new firmware, notifying the node when it is complete. The updated node then restarts using the new firmware.

Details of the requirements for the node and the CT can be found in the Firmware Upgrade Standard. Most users will not need to know how the derails work, but using the Firmware Upgrade will be important. New features have been distributed and errors corrected using this process.

# Network Implementation

As noted earlier, LCC does not require a specific network to operate. Although the CAN™ bus is the only network currently represented in the standards, other networks have been considered and may be defined in the future. WiFi networks would be convenient for installation and are actively being pursued. The possibility of connecting to other existing networks like LocoNet™ or CMRI (Computer Model Railroad Interface) has also been considered, and some hobbyists are building these interfaces. The hobbyist that just wants to install an LCC network only needs to know how to connect the nodes. An overview of the network considerations is provided for basic understanding of the networks and the interfaces that support them.

Some of the considerations in choosing a network for implementation come from the goals of the LCC. LCC is a distributed network with no master computer to direct traffic, so the network needs to be a broadcast form where every Node receives all the messages on the network. In the descriptions of messages above, the underlying assumption was that a broadcast network was in use.

Another assumption built into the LCC messages is that the network transfer mechanism provides error detection and collision avoidance. (Collision avoidance means that a node cannot pass a message to the network at the same time as another node.) Neither of these functions are included in the messages defined by LCC.

## CAN Bus

The CAN Bus is probably the most used network for LCC at this time. CAN was chosen as a hard-wired network because it is a mature technology which is used in automobiles extensively. This means that it has a complete range of interface integrated circuits available at a low cost.

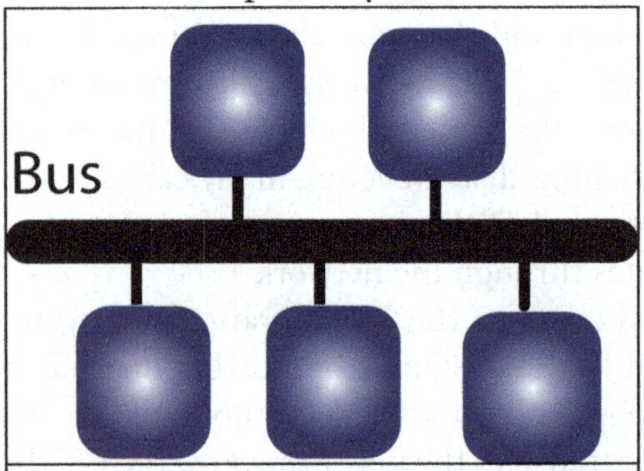


**Figure 1.6 A Bus Network connects all Nodes to the Bus**

The CAN Bus provides message transport with error detection and collision avoidance. These functions are implemented at the interface level and the LCC software does not worry about either one. CAN is also a broadcast network which connects to a single cable. This is similar to the way multiple locomotives can connect to the track, which creates a bus (cable) to deliver data to all the locomotives. Since CAN is a serial network, it can be implemented with only three wires. In LCC, this is accomplished by using low-cost cables widely used for interconnecting computers. The cables are the CAT5 cables available everywhere. The topology of a Bus network is shown in Figure 1.6.

While the CAN Bus has most of the features that LCC needs, it does have some restrictions, especially in the size of the messages. Because the message header of a CAN message is shorter than the defined message header of LCC, a modified message headers is required. The primary change is to shorten the Node Identification by obtaining an alias at the time of Node Initialization. This alias is then used for messages on the CAN network. When the full Node Identification is needed, requests for the Node Identification can be sent to the network. EventIDs are not affected by this alias, because they are not necessarily associated with a Node Identification and an alias could not replace the first bytes.

Another restriction imposed due to the CAN Bus is the length and termination of the cables. Termination of the cables is required to match the impedance of the cable. Commonly this is either built into a Node, or is available as a separate plug-in device. One termination must be on each end of the cable. The total length of the cable depends on how many physical node connections are on the cable.

## Other Possible Networks

Wireless networks are popular in all communications. The most common network after the Cellular Phone network is the WiFi network commonly used to connect to the Internet. Anther use of WiFi is train control, either with dedicated transmitters and receivers, or to a specific software such as JMRI. In both these cases, the network connects to a specific software to control or access the functions defined. These connections are called Point-to-Point connections, because they connect one (or many) nodes to one specific Node called an Access Point. In the network messages a single address is used to communicate between the Node and the Access Point.

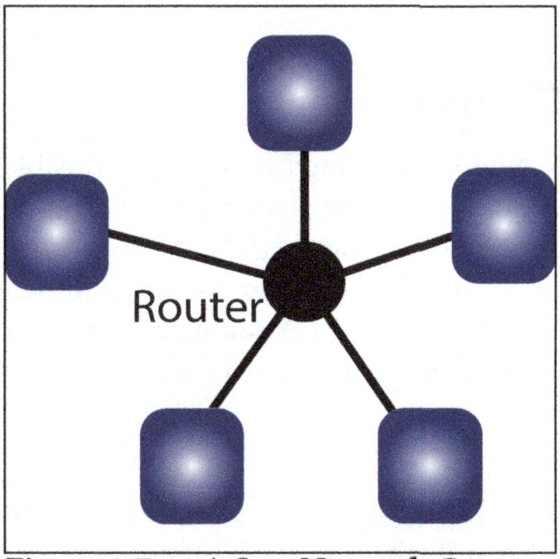


**Figure 1.7 A Star Network Communicates through a central Router**

A generalization of this network topology is known as the Star Network (The Access Point topology with a many-to-one node is also a Star Network.) In a Star Network, a central device known as a Router receives all messages from the Nodes and then "Routes" the message based on the address contained in the message. This topology is shown in Figure 1.7. In a Star network, all nodes send messages to the Router in the center. The Router then decides where to send each message. To be useful for LCC, either the Router has to send the message to all nodes, or it has to have a table defining which Nodes want which Messages. Since some of the messages apply to all Nodes, sometimes the Router has to send a message to everyone. This can take time, since each Node is getting a copy of the message. Another disadvantage of the Star topology for LCC is that the Router is a single point of failure. If the Router fails, the entire network is down. This network topology would be used with the wireless implementation of LCC. The simplest implementation would be to create a Router that simply copied all messages to all Nodes. The disadvantages of a single point failure and the added cost for a high-performance Router make this a less desirable network at this time.

A popular form of the Bus network is created with a Master Node to send messages to all Nodes as shown in Figure 1.8. In this network, the Nodes depend on the Master to distribute messages. In most cases, the Master has most or all of the data processing and the Nodes are simple sensors or outputs. While the bus can be simple, the Nodes do not have computational ability to act on their own. The single point of failure is the Master Computer. Since there is little or no processing in the Nodes, the processing time is serialized on the Master computer, therefore, the Master may limit the total throughput of the system. Because of these disadvantages, LCC did not choose this network form.

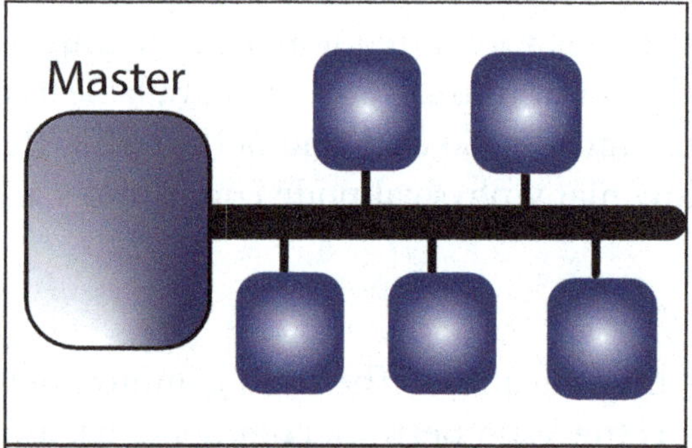


**Figure 1.8 A Master-Slave Network Depends on a Master**

# Inter-Network Connections

A CAN Segment is a set of CAN Nodes connected to a common cable. When the cable length gets too long, or a different type of network needs to be connected, some inter-network connection needs to be used. If the connection is to a computer port (USB) then the device is called a buffer. This section will describe these devices and show how they differ. The descriptions are based on CAN LCC bus, but could be extended to other networks. The primary network will always be an LCC network.

## USB Buffers

A USB Buffer is an interface between the CAN Segment and a computer. The device simply buffers CAN messages and transmits them to the USB port and vice versa. The USB signal is connected to a computer port to provide LCC messages to programs inside the computer. Additional functions of the buffer may include electrical isolation and power/activity indicators. In use, the USB Buffer counts as a load on the CAN network. This is the primary device needed to connect to a general purpose computer for configuration. Both current Configuration Tools use a USB Buffer for network connections.

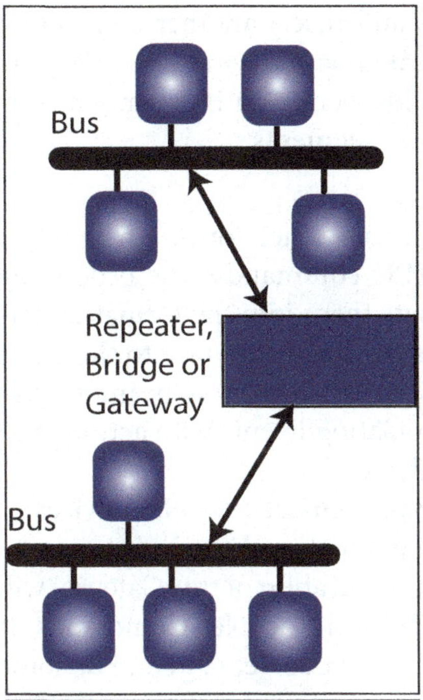


**Figure 1.9 Repeaters, Bridges and Gateways connect networks**

## Repeaters

The box in the center of Figure 1.9 can represent one of the three listed devices. Perhaps the easiest device of this set is the Repeater. A Repeater connects two CAN Segments by copying all activity on one segment onto the other segment. This is done on the bit level, so no knowledge of the message type, source, or destination is required. The segments look as if they were one contiguous segment. The purpose of a Repeater is to extend the network beyond the physical length restrictions imposed by cable and interface losses. To accomplish this, Repeaters will add a very small time delay to the messages. Due to the constraints on timing in the segment, this time delay slightly reduces the maximum length of the segment. At this time, at least one repeater is commercially available.

## Bridges

A Bridge is another way to connect two LCC Segments. In a Bridge, the entire message is received and resent as required. A Bridge appears as a Node on both networks and is addressable on both networks. The two segments connected to a Bridge operate independently of each other. Both segments run as if the other segment is not there. Bridges need to know which messages are needed to pass on to the other segment. In a CAN segment, the process of assigning aliases occurs on each segment, therefore the source of a message from one network becomes the Bridge itself. This means that the Node IDs on the segments do not know each other — only the Bridge knows how to relate IDs on one segment to the other.

Since the Bridge needs to analyze and modify the messages, a longer time is required to pass through a bridge. The time delay is only significant to the use of the message, but does not affect the length of the segment. Two or more full-length segments could be connected with Bridges.

Because the time delay does not impact the LCC message, a possible use of a Bridge would be to split the CAN Nodes and insert another connection in between. For example, a Bridge could be created with a wireless connection between two CAN nodes. The Can Nodes would operate on independent segments and allow large distances between thee two segments.

### Gateways

A Gateway translates LCC messages to some other form. An example would be a Gateway to translate LCC into CMRI commands. The processing in a Gateway receives the messages and reformats them into the format understood by the other network. Translation of messages and interface to the other network are required. Gateways are very specific to the network to be interfaced. They also may be installation specific, since associating Events with actions and sensors may require knowledge of the specific devices.

Of particular interest may be a Gateway to connect LocoNet devices to LCC. In this case, the LocoNet commands are defined, and the Events can be mapped to specific LocoNet commands. User configuration of this Gateway will require a table of associations between ***EventID***s and LocoNet Commands. It may be tedious, but the result will be a hybrid LCC-LocoNet system. For some cases, this will allow expansion into LCC.

## Future Expansion of LCC

New messages, Event types and network implementations are in-work. Additional manufacturers are designing boards. New applications pop up all the time. LCC is not a static system. As new technology comes available, LCC can adapt.

Experimenting with wireless networks is a popular topic. WiFi networks are inherently easier to create, but have problems of their own. Current wireless protocols are primarily peer to peer networks and do not transmit all messages to all users.

An interesting application might be a turntable interface. Entering a stall number for a turntable would move the table to the requested stall. This would aid automated operations.

There has been some discussion about an event with a "payload" attached. The idea is that some events need more than just reporting the occurrence of some change. The result of the change may also need to be reported.

## Summary of LCC Basics

LCC is a distributed network to control the model railroad layout, not the trains. LCC augments and complements the DCC system by controlling accessories such as turnouts, signals and scenery. LCC uses the Producer-Consumer model to identify interactions between LCC Nodes. These interactions, the

Events of the network, represent state changes in the layout. Events are passed from Producers to Consumers to cause an action at the site of the Consumer.

LCC defines identifiers for the Nodes and Events to be Globally Unique. This means that the Nodes can be safely added to any network without causing conflicts with other Nodes. Node IDs are assigned at manufacture based on ID ranges that are assigned to the developer of the Node.

Since the network is distributed, physical access to the Nodes is not guaranteed. To allow configuration of the Nodes over the network, Configuration Data is embedded in each Node. The Configuration of a node is the process of associating Events with I/O on the Node. Configuration Tools on computers provide a human readable process to set the variables in the Configuration data.

The LCC network is not specific to any network implementation, but it has been implemented with the CAN bus. This is a mature stable technology in use by the automotive industry,. The CAN Bus provides reliable message transfer for the LCC. Other network implementations may be possible in the future.

Details of the message bits are not necessary to assemble a working LCC system. Only developers need to read and understand the standards. The following chapters will attempt to guide you in the creation and construction of an LCC system. Starting with a simple LCC network, the construction of a demo layout will demonstrate the Configuration and logic required.

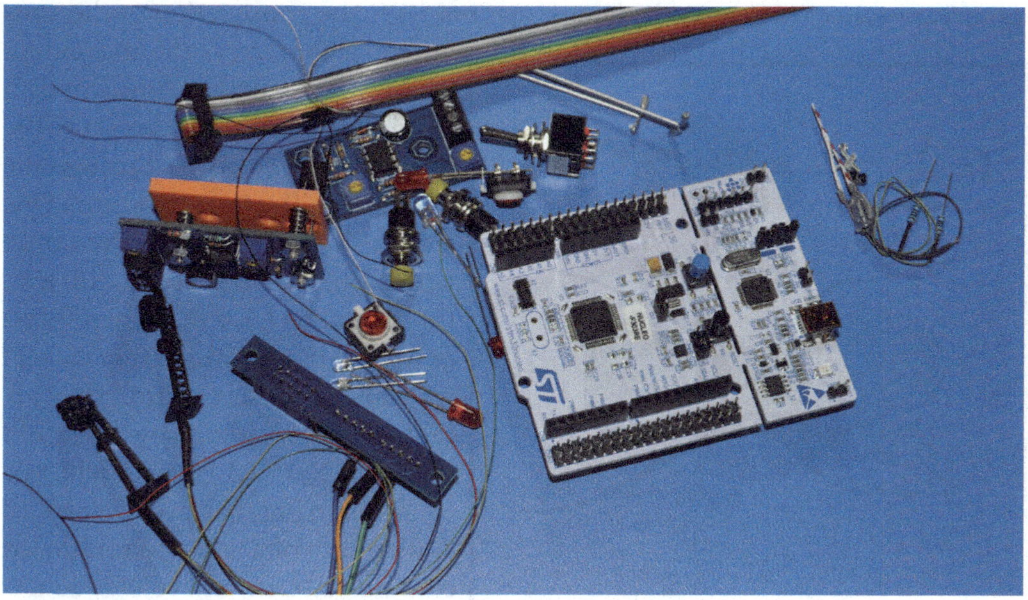

# Chapter 2 – Parts and I/O

Now that you have an idea about LCC, its terms and concepts, what do you do with it? Connect up a bunch of parts and see how it works? Of course, that is the purpose of LCC, to connect parts to control the layout. Okay, but how do you do that, and what are those parts anyway? Well, first things first. (Unless you already know all about parts and interfaces. If so, just skim this chapter for the highlights.) First, we need to define some basic properties of the parts that we put on a model railroad. So, this chapter will start with the electrical properties in general, then see how some of the common parts on a layout work, and then cover the interface to common Node I/O pins.

## Electrical Properties

Electrical Engineers toss around terms like Ohms, Volts, and milliamps expecting everyone to know what they mean. I know, because I am one. These basic electrical terms have been defined forever (well maybe a couple of hundred years.) Most of the hobbyists reading this are also familiar with these terms, but may not know how they relate to each other. The answer is Ohm's Law, one of the most fundamental laws about electricity. After Ohm's Law, two laws attributed to Kirchoff define how parts of an electrical circuit interact. These simple laws are all that the ordinary hobbyist needs to understand how to put together an LCC System.

To illustrate Ohm's law, the cartoon in Figure 2.1 shows the parts of the law. Voltage (in Volts, symbol V) is considered the potential (shown as a pressure) for moving something (electrons) through a conductor, shown here as a pipe. Resistance (measured in Ohms, symbol R) is the constriction of the conductor which slows the movement of the item. Current (measured in Amperes, symbol I) represents the movement of electrons through the conductor. The form of Ohm's Law that is shown here is in the upper Left corner of Figure 2.1. In text, this would be I = V/R. In many cases we will be looking for the correct resistance to insert to limit the current to some value, so we will use the form R = V/I (rearranging the equation using that pesky algebra.) Sources of voltage are common, including batteries and voltage adapters. Resistors, incandescent light bulbs, and motors provide resistance. These are simple parts that are common in model railroads. Other parts are not as simple. An example which many don't understand is the Light Emitting Diode (LED) The care and feeding of LEDs will be covered shortly. First, we need to look at Kirchoff's Laws of circuits. The circuits in Figure 2.2 illustrate Kirchoff's laws. In

**Figure 2.1 Ohm's law relates Voltage (Pressure), Current (Movement speed) and Resistance (Constriction)**

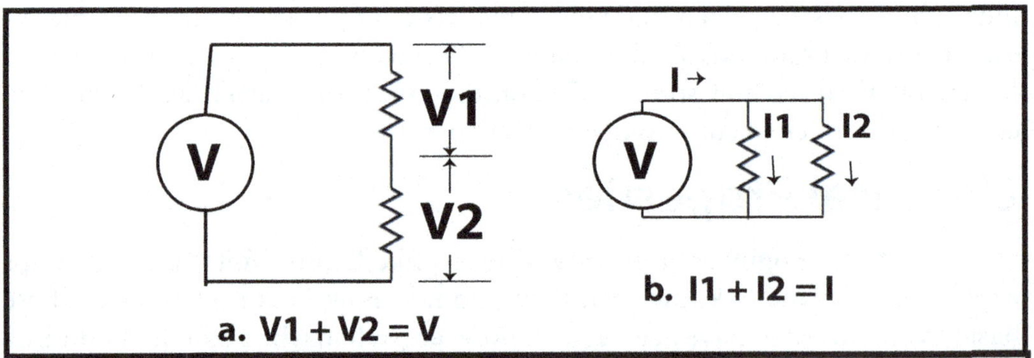


**Figure 2.2 Kirchoff's Laws determine the Voltages and Currents in Complex Circuits**

these drawings, the Resistor symbol is that zig-zag thing. A voltage source is represented by the circle with a V in it. Stated in words, **circuit a** states that the sum of the voltages in a series circuit equals the total voltage. (This drawing represents a series circuit.) Similarly, **circuit b** states that the sum of the currents in the parts of a parallel circuit equal the total current from the voltage source.

Okay, that seems obvious, why did it need to be stated? Because it needs to be considered whenever circuits have more than one part. Remember **circuit a** in Figure 2.2 when you hook up an LED, and **circuit b** when you add another device to the power supply.

One more electrical property that is important in some cases is power. Power, measured in Watts, is a source of heat and can cause parts to burnout. Burnout is a good term, because too much power heats up a part and can melt wires or cause internal damage. This is colloquially known as "Letting the smoke out." When we use a part, we must be aware of its power rating. This is especially true for resistors used to drop the voltage supplied to another part. Fortunately, resistors are available in many power ratings. Power in Watts is the product of Volts and Amps, P = V x I. So, 1Watt is defined as 1 Volt times 1 Ampere.

Additional electrical concerns will be detailed below. I must warn you that not all circuits are as easy as these. Ohm's law has to be extended to apply to alternating currents or pulses. While the basic form is the same, the devices that are common in alternating current parts do not have a constant resistance. In most cases, these concerns do not impact the installation of an LCC system.

Electronic symbols will be used in the drawings below. As I use them, I will point out what the symbol represents. If you search for electrical symbols, you will find many — more than I need to use here.

# Some Basic Parts

This section will introduce some of the very basic parts that are seen on model railroads. These parts are normally the smallest parts and are either soldered into the circuits or held in terminal strips. I will not attempt to describe the details of how these parts work, only as much as is needed to use them. These basic parts come in three categories: switches, indicators or lights and actuators.

## Switches

Basically, the purpose of a switch is to connect two (or more) wires. The switches in Figure 2.3 are typical of switches used with LCC Nodes. Two types are shown, a toggle switch and several pushbutton switches. As shown by the symbols, the switch connects two or more wires together, allowing current to flow through the switch.

The DPDT switch shown has "Double Poles, Double Throw" action. This is the way switches are identified. It means that two circuits are connected at the same time and that there are two different circuits to connect. In this naming convention, a SPST switch would connect a single circuit to only a single circuit.

The term "toggle" is a name for the "handle" that you use to change the switch. Other types of "handles" are slides and rockers. These are classified as DPDT, SPDT, etc. in the same way as toggle switches.

The pushbutton switches at the bottom are momentary switches because they are only connected while they are pressed. The symbol for the pushbutton shows this action. Other variants of pushbutton switches are possible including multiple poles, "normally closed" that open the circuit when pressed and "push-on/push-off" which alternate between closing the circuit and opening it.

**Figure 2.3 Typical Switches**

Switches are usually simple to wire, because the wires are attached to the terminals or pins on the switch. The only tricky thing about these switches is that they are often opposite of the direction you think. A toggle switch is a lever that pushes a contact back and forth. That means the toggle portion points in the opposite direction of the terminals that are closed. Rocker switches are similar in that they are usually closed on the terminals opposite to the rocker side that is down. Slide switches Operate in the opposite way — the slide is in the direction of the terminals that are connected.

## Indicators and Lights

Indicators and lights are all over the model railroad layout. By far the most popular indicator at present is the Light-Emitting Diode (LED). Several parts shown in Figure 2.4 are LEDs. In some cases, you may find an incandescent lamp still used for lighting. Finally, there are some special lighting effects like electroluminescent panels that require special boards to operate. These parts may only allow switching on and off. Specialized interfaces will be required for this class of part.

**Incandescent lamps:** Lamps that have a filament are similar to a simple resistor. They have two leads (or are in a socket with terminals) and polarity does not matter. (Polarity means that the leads must be connected with one specified as positive and one as negative.) Incandescent lamps have voltage and current ratings which limits the input voltage. Since they are effectively a resistor, these lamps obey Ohm"s Law. If a resistor is desired to drop the voltage or the current, the law of series circuits can guide the determination of the resistor value. Incandescent lamps emit heat when they are lit. This heat needs to be vented, or the surrounding parts can become damaged. Also, when an incandescent lamp first turns on, the filament is not hot, and it has a lower resistance. This results in

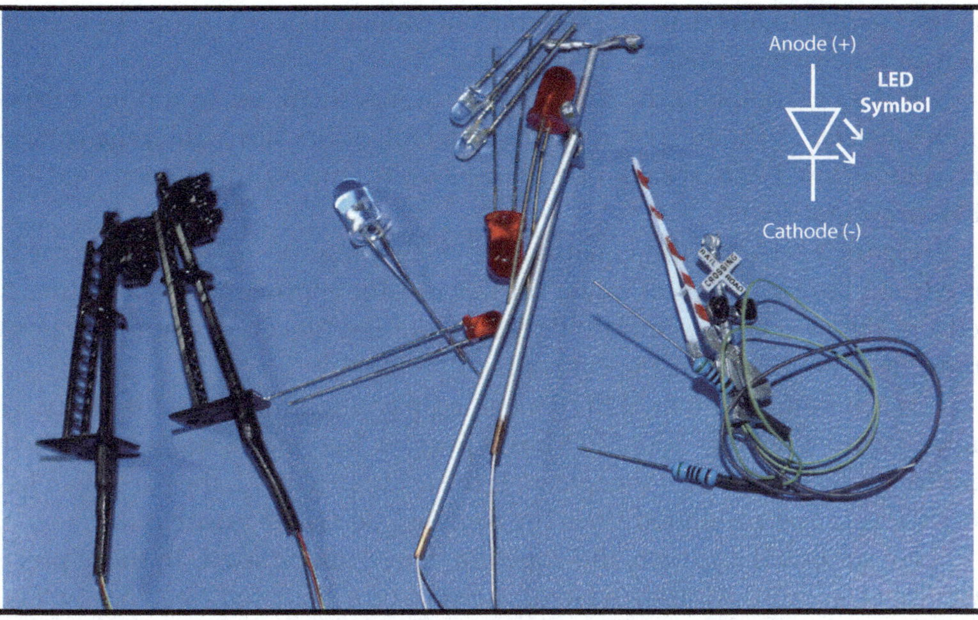


Figure 2.4 LEDs are Popular for all Model Railroad Lights

an initial current that is higher than the steady-state current. You might have to consider this when deciding on a power supply for this purpose.

**Light Emitting Diodes (LEDs):** LEDs are now available in all types of lighting. From the small LEDs in the headlights of model locomotives to the headlights of full-scale locomotives, LEDs have become the light type to use. LEDs now come in multiple colors, sometimes even within the same package. These colored LEDs are very convenient for model railroad signals, including searchlight (all three colors from one package) or color light signals (three independent packages for Green, Yellow, Red.) White or slightly yellow LEDs are used for locomotive headlights, street lighting and internal structure lights.

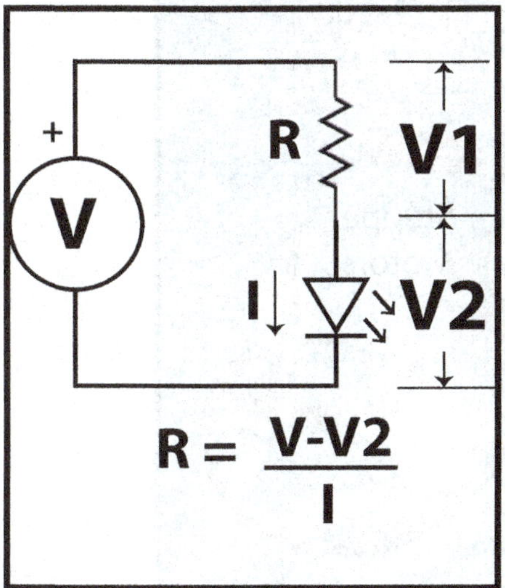


Figure 2.5 Calculating the Current Limiting Resistor

An LED is different from an incandescent lamp — an LED is essentially a constant voltage device. This means that the voltage has to reach a certain level before the LED lights, and it does not increase in voltage much above this amount. Current does increase and determines the brightness of the LED, but has an upper limit. Only in very rare cases should you use an LED without a current limiting resister in series with it. Most of the time, limiting the current to less than 10 milliamps (a milliamp or ma. is one thousandth of an amp.) provides adequate brightness.

If the LEDs all had the same voltage across them, we would be finished. Unfortunately, LEDs are not so nice, Different colors have different voltages. Red and Yellow LEDs seem to be about 2 volts, and White, Green, and Blue LEDs are about 3 volts.

Figure 2.5 shows the equation for calculating the resister value needed for current limiting. Typical values for this resistor approximately 300 ohms for Red and Yellow and 200 ohms for Green, White and Blue. Calculating the power rating gives a number less than ⅛ watt.

When connecting an LED you must pay attention to the polarity. As shown on the drawings, the LED has a positive lead and a negative lead, The positive lead is always the longer lead. If the leads are cut off, the base of an LED has a flat side and a rounded side. The rounded side is the positive side.

For an initial estimate, the equation above can be used. In more precise cases (not common) additional factors need yo be considered. Specifically, when the LED is controlled by a microcomputer, the voltages need to be adjusted. These considerations will be shown in the later section on interfacing microcomputers.

## Actuators

Actuators make something move. The most common actuator is the motor. Every layout has some motors. LCC is very good at causing a change in the state of

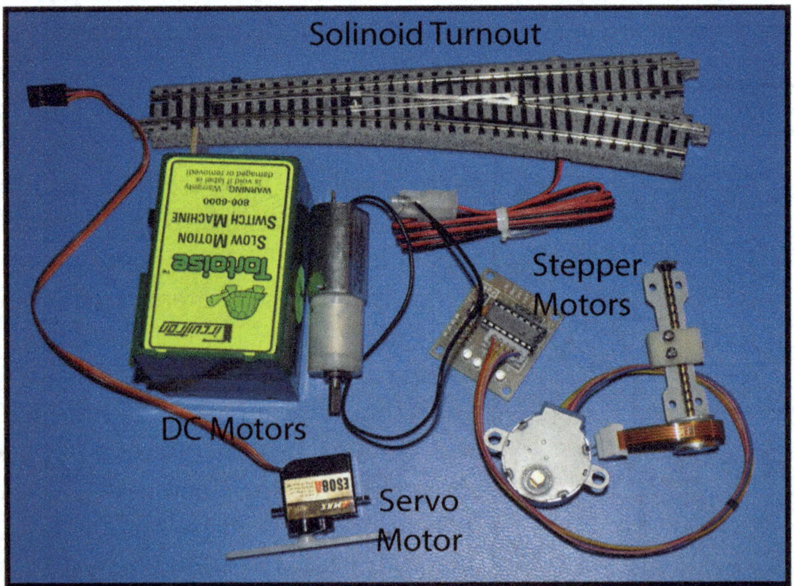


**Figure 2.6 Actuators provide Motion on a Model Railroad**

something, so that implies motion in many cases. Figure 2.6 shows some popular actuators.

**DC Motors:** Two DC Motors are shown in Figure 2.6, although one is hidden inside the Tortoise™ turnout machine. The cylindrical motor is typical of a geared motor that could be used for animation on the layout. Both motors are simple to connect — Just two wires. Polarity determines the direction of the motor. As long as you only want to go one direction, that is all that needs to be wired. To provide reversing and speed control, the normal connection of DC motors is through a device known as an H-bridge. To understand the function of an H-bridge, an H-bridge is made up using four switches to connect to the motor. These switches can be part of a mechanical switch — a DPDT switch — or semiconductor switches allow speed control because they can be switched on and off to perform Pulse Width Modulation. That is the common form of speed control used in DCC decoders. To understand the H-bridge, refer to Figure 2.7. In the figure, switches **a** and **d** are currently closed, allowing the positive voltage to connect through the left motor terminal to the right. If switches **b** and **c** are closed and **a** and **d** are open, the motor voltage will be reversed and the motor will run in the opposite direction.

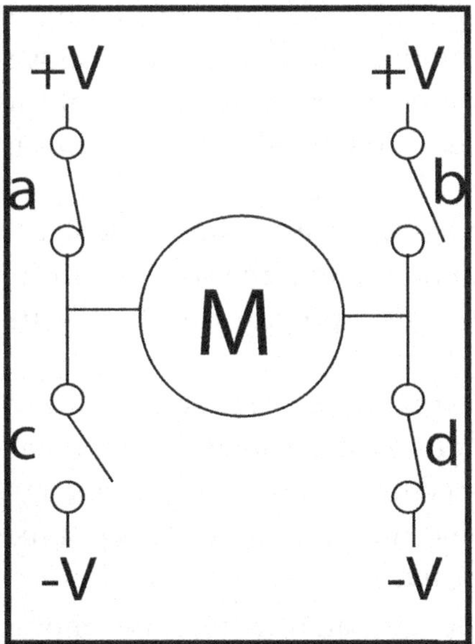


**Figure 2.7 Four switches make up the H-Bridge**

H-bridge outputs are not referenced to the negative side of the voltage supply. Unlike most outputs, the two wires from an H-bridge switch between the positive voltage and the negative voltage.

As shown in the figure, there is a possibility that two switches on the same side of the bridge are closed at the same time. To prevent this, most switches are opened before they are closed, ensuring that the +V and -V are not connected at the same time.

Control of "Stall Motor" turnout machines use some form of H-Bridge to control the direction. The Tortoise motor in Figure 2.6 is typical of this type of actuator. Stall motor machines are named for the fact that they only move a short distance until they hit a barrier that prevents further movement. The motor then stalls and keeps pressing on the barrier. Because the current is low, the motor does not burn out and it keeps pressure on the turnout.

**Solenoid Turnout Machine:** Another use for the H-bridge is the single coil solenoid turnout. These turnouts use a single coil solenoid to throw the turnout points. (A solenoid is simply a magnetic coil that pulls or pushes a iron rod to move it.) In the single coil turnout machine, the voltage has to be reversed to change the direction of the turnout. The only major difference in the single-coil solenoid and the stall motor is that the solenoid takes more current. The heat build-up would melt the plastic, so these turnouts need to be pulsed —turned off shortly after turning on.

Other forms of solenoid actuators include the dual-coil turnout machine and simple solenoids for linear motion. These both operate with a single switch. These may need the pulsed voltage like the single-coil solenoid or they stay energized. Current requirements in these cases can get very large.

Another form of the solenoid is the mechanical relay. The relay has switch contacts on an arm which is pulled by a magnetic coil. The advantage of the relay is that it can be operated by a low current device (the LCC Node) and control high currents. The disadvantage is that it is mechanical, and can wear out. Dirty contacts can also be a problem. As technology advances solid-state switches will be more common.

**Stepper Motors:** Two stepper motors are shown in Figure 2.6. Stepper motors are primarily used for positioning or precise speed control. A stepper motor actually moves one step at a time. A step is a fraction of a revolution. If the steps are fast enough, the motion seems smooth and can be used for movement. Stepper motors require special control which is beyond the scope of this chapter. To control stepper motors, software must sequentially energize coils in the motor.

In Figure 2.6, the larger stepper motor has a circuit board attached to provide the current needed for the windings in the motor. The software in the controlling computer will interface with this board to provide the steps at the correct times.

The other stepper motor is a linear actuator which drives a screw to move a sliding section. This is very much like the designs used in cameras to automatically adjust focus. If you want a precise positioned linear motion this is one way to go. Notice that the connector on this motor has multiple conductors just like the other stepper motor. It needs the same software.

**Servo motors:** A popular device in model airplanes is the servo motor. This motor adjust a lever arm to an angular position based on the voltage on a control line. Three wires are connected to the servo motor, two for power, and one (reference to the negative wire) for the control voltage. This control voltage can be a PWM (Pulse Width Modulation) signal, so a PWM output can control a servo motor.

PWM is one of the common outputs from microcomputers. As LCC nodes are developed, this will become a viable option for actuating turnouts, crossing gates, and semaphore signals. Since servo motors are relatively inexpensive, they are likely to be very popular.

# Block Occupancy — What is it?

One of the requirements for implementing signaling is Block Occupancy. Signaling needs to know if a signal block is occupied to select the proper signal ahead of the block. A later chapter will discuss the LCC capabilities to implement signaling. Here I need to discuss types of block detection and how they affect operations.

First, there are two broad classes of Block Detection — those that connect to the track and those that do not. There are advantages and disadvantages to both classes. This chapter introduces these detectors for interface information. Chapter 8 will look at more details.

## Current Detectors

Detectors that connect to the track operate by detecting current in the track power, therefore implying that there is something on the track. Obviously a locomotive will require current to run, so the locomotive will be detected. If a car has lights it will require current. Cars that do not require current will be invisible to this detector. The solution is a small resistor connected between the rails. Usually this is accomplished by adding a resistor to a wheel set. With current technology, these resistors are so small you almost cannot see them.

The question then is: what are the characteristics of these detectors that we need to consider? Popular detectors are designed either with a small voltage drop or with a current transformer. Without going into details, the voltage drop is detected because there is current passing through some part, usually a diode pair connected in opposite directions. The voltage drop can be significant but generally is around 1 volt or less. While this seems low, when a locomotive transitions from a block with a detector to one without a detector, the speed seems to change. This is especially true for DC locomotives. Yes, this detector class works on both DC and DCC. However, when used on DC the detector will not work unless a small voltage is applied.

A current transformer is a simple transformer that detects current by coupling the current in a wire passing through the transformer to coils wound around the same core. There is no wired connection between the track power and the detector. Unfortunately, current transformers only work with alternating current such as DCC. The advantage is that the track power is not changed by the current transformer. DCC provides another advantage because it always has voltage on the rails. Therefore, current detectors with DCC are operational no matter how fast the locomotive is going.

## Optical Detection

Detecting the interruption or presence of a beam of light also provides a method to detect something. The normal way to make these detectors is to mount a light detector between the rails. When a train passes over the sensor, the sensor detects the interruption of the light and signals that. Because this works on the interruption of a light source, either the detection does not work in the dark, or a source of light must be provided.

This class of detector is considered a "Check-in, Check-out" detector. That means that a block must have a sensor on each end, one to indicate the block is occupied and one to say it isn't. This is often a problem, for example, if a locomotive is removed from the block without going through the second detector. Also, the scheme is direction dependent, so you need to know which detector is triggered first, then wait for the other detector. Backing out of the block is another concern that needs to be considered.

Most likely, current sensing is best for block occupancy detection. In some cases, optical detection will be okay, especially when the detection is limited in length. Using optical detection for a speed measurement or to stop a train at a identifiable point are good applications.

### Mixed Block Detection

Recently a mixed form of occupancy detection has been implemented. Using the best capabilities of both current and optical detection, a block can detect trains without requiring current to flow through every car. Since trains on the mainline are always connected to a locomotive, the current through the locomotive will activate the current sensor as long as the locomotive is in the block. Adding an optical sensor at the boundary of the block will sense cars still leaving the block. The combination of these two provide a block occupancy which does not require adding resistors to cars.

## Common Interfaces Out and In

The basic parts described above are inputs and outputs for more complex devices such as microcomputers and LCC Nodes. To correctly interface with the more complex devices you need to know some of the basic characteristics of their outputs and inputs. In some cases, the output of a "more complex device" can be an input to another device. A simple detector for block occupancy is an example. Figure 2.8 shows a few of these more complex devices. As technology has advanced, inputs and outputs of these devices have become somewhat standard-

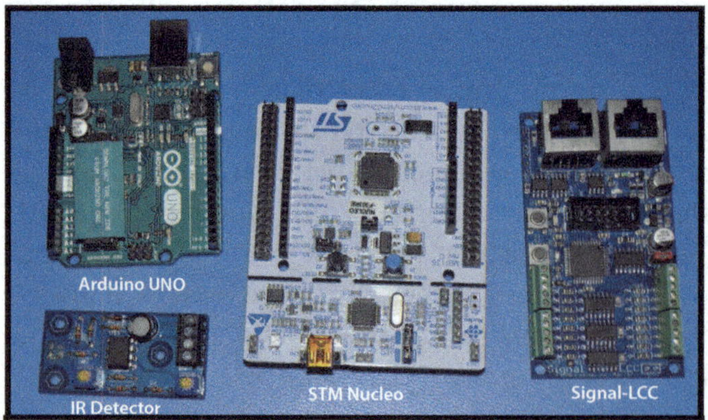


**Figure 2.8 Typical Boards for Model Railroads**

ized in their function. It is those functional forms that will be described below. Recognizing that the devices are Integrated Circuits (ICs) with transistors as their primary input or output, some of the names are related to transistors. To better understand what they mean, I will use switches and resistors to model the inputs and outputs.

### Isolation

Electrical isolation is the process of separating devices from each other so that no current flows between the devices. In most cases, the devices need to communicate with each other, so some form of interface is required. When devices are powered by different sources, the reference to "ground" can be ambiguous. Interfacing cannot depend on the voltage levels, because current flowing between the devices may raise or lower the "Ground" reference. Isolation devices can establish a known interface between devices such as sensors and microcomputers.

Transformers are traditionally the most popular type of isolation. Transformers are a well-known technology and are commonly used in alternating current devices such as power supplies. Transformers are based on magnetic flux changing and therefore do not work well on DC. On AC (even on low spreed square waves) transformers can also change the voltage. Their use in the line power in our buildings is an example. However, do not be misguided into thinking that all power supplies plugged into the wall are isolated, because some are not.

A popular use for model railroads is the use in occupancy detection. In this case, the track power is detected by a "current transformer." In this case, one side of the track power is run through the transformer. This transformer provides both a step-up and an isolation. Because this form of detection relies on transformers, it is not applicable for strictly DC systems

A simple relay can also provide isolation and voltage changes. Activating the relay requires a voltage and current that may be difficult to get from sensors, but the relay contacts are generally totally separate from the coils. Since relays are magnetic, it takes a little time to build up the field and then a little longer to make the mechanical movement. Because of this time delay, relays are best used on slow changes.

Optical isolators are the latest form of isolation, Based on transistors, an opto-isolator combines an LED and a transistor in a common package. When the LED illuminates, it shines on a photo sensitive transistor, causing it to turn on. Different types of transistor can be used, depending on the power and type of control needed. The input to these opto-isolators is like the LED discussed earlier. Usually a low voltage (maybe 1.5 volts) and a low current (usually less than 5 ma.) Outputs are like the Output Models discussed below.

Since opto-isolators are based on the same technology as the devices we interface with, the task of interfacing is much easier. Opto-isolators are fast (working at light speed), work well with DC, and work at the voltages we are using.

Another form of opto-isolator is the solid-state-relay. In essence, a solid-state-relay (SSR) has a higher current and voltage rating, so it can be used in more power circuits. These relays may also have heat sink capabilities to control even more power.

## Common Output Models

Outputs from these devices fall into one of three types with two variants. The first variant is implemented with transistors(aka junction transistors.) This was the first type of output when ICs were first designed. Models of the transistor type are shown in Figure 2.9. The most common of these is the Common Collector type. The Common Collector output is not connected when it is "Off" and connected to the negative voltage side (often referred to as "Ground" or "Grounded") when "On". similar to the LED, transistors have a small constant voltage offset when they are conducting current. This is represented in these models by a circle with a V in it. Note that when the Common Collector is Off, the output is not connected to any part of the circuit. This is referred to as a "floating output" because it is determined by the other circuits it is connected to.

Similarly, the Common Emitter type is connected to the positive voltage side of the voltage source. The small voltage offset still appears in this case. A major difference between these two type is that the Common Collector is not limited to the voltage of the IC. Frequently, a Common Collector output switches higher voltages than the IC voltage. This is convenient in interfacing this type to other circuits.

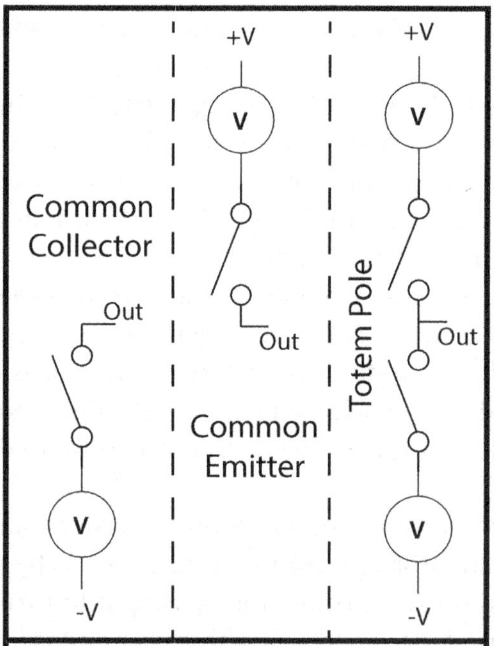


**Figure 2.9 Transistor Outputs have a Voltage Offset**

The third type shown is called a "Totem Pole" output because it looks like a Totem Pole. This is a combination of the previous two types. It has the advantage that it does not have a floating output. Current flows from it or into it depending on whether it is Off or On. As with the other two types, the transistor variant never quite reaches the full limits of the voltage source.

The second variant of output uses a Field Effect Transistor (FET) instead of the common junction transistor. FETs have desirable features for this purpose. First, instead of a voltage offset, they look like a small resistor in series with the

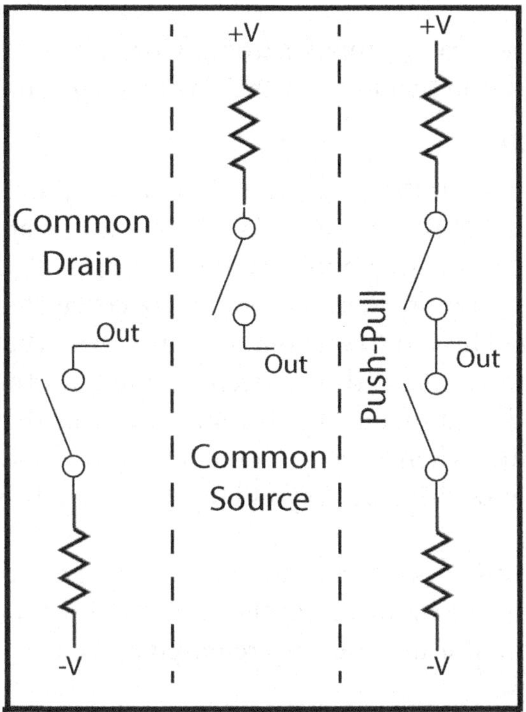


**Figure 2.10 FET Outputs have a very Small Resistance when On**

switch. This is shown in Figure 2.10. Since this resistance can be less than an Ohm, the resulting voltage offset is small. The second advantage of the FET is that they are typically higher current devices. Referring to v, the same types are seen, with different names. The names "Drain" and "Source" are names used in FET symbols, so they are applied in this case. Since the resistance in a FET is small, the "Push-Pull" type is much closer to the limits than with transistors. Also, the current available is greater. The problem (or is it an advantage ...) of the floating outputs is still present. The next section on Inputs will explain the advantages and disadvantages of "Open" outputs.

## Common Input Models

The most common input model is simply a resistor to Ground (the negative side of the voltage). In digital circuits, most times you can ignore any other characteristics of inputs. Input resistance is also normally high. That means the current into the device will be small.

The other characteristic of inputs is the input level must be above a specific voltage to be "On" and it must be below another level to ensure an "Off" condition Usually that is defined with a region between the "On" and "Off" where the input is not defined. For example, if the Arduino™ UNO shown in Figure 2.8 is powered by a 5 volt supply, it will read an input as Off when the input voltage is less than 1.5 volts, and On when the input voltage is greater than 3 volts. The 1.5 volts in between are not guaranteed to be either. If the input does not firmly set the On and Off voltages, the circuit may not work correctly. All simple inputs have this characteristic.

The problem with this characteristic is that it is not tolerant to the floating outputs above. The solution is simple — just add a pull-up resistor to the input. A pull-up resistor is just what it sounds like, a resistor connected between the input and the positive voltage. Remembering that the input current is very low, this resistor can be relatively large, maybe 10,000 ohms or more. Fortunately, this is usually not necessary, because many of these devices already have pull-up resistors built-in. In the case of most microcomputers, the pull-up resistor needs to be turned on. The RR-CirKits Nodes turn on the pull-up resistors when an I/O is configured as input.

Pull-up resistors are necessary whenever an input is from a Common Collector, Common Drain or from a pushbutton that just shorts the wire to Ground.

### Combined Input and Output

In the network case, input and output are on the same wire. Several serial networks operate this way, including the CAN Bus for LCC. While the details are technical, the basic principle relies on that floating output of the Common Collector or Common Drain type of output. If more than one output are connected to the same wire, the output would stay High until one of them turns On. That one output would pull the wire down to Ground. If there were an input on the wire, it would see that Low voltage and read that as input from the device that pulled the wire Low. Networks determine when they can send data by the state of the wire. The methods of making that decision are beyond this book, but this is the basis for networks.

The Open Collector network described still needs some way to make sure the High level is high enough. In one of the inputs, or at the end of the wire, a pull-up resistor is added. This is a terminator like the ones required by LCC.

## Interfacing to a Microcomputer

In LCC each node is a microcomputer. Often the interface is provided by another board connected to the node, but many applications require a direct connection to the microcomputer. Microcomputers (μC) used in LCC are general purpose μCs which are a single Integrated Circuit, usually in a multi-pin package. This type of μC has what are known as General Purpose Input/Output (GPIO) pins which can be used for either input or output. This function is controlled by the program in the μC.

Development of the Arduino system has enabled the personal use of the μC to many hobbyists. Model Railroad applications have been created for multiple purposes. Because the Arduino system is user-friendly, many people can now program μCs for unique purposes.

The heart of the Arduino system is a hardware μC combined with an Integrated Development Environment (IDE) which provides a program editor and automated process for compiling, loading and executing the program. Many libraries of programs are available on-line to provide examples and starting points. Since the inception of Arduino, many μCs have been added to the IDE. Very fast and multi-functional μCs are available, including some with CAN support. As the market for Arduino grew, add-on boards were produced to provide special interfaces. Further details about the software and hardware of Arduino are beyond the scope of this book.

The Arduino μC is typical of the μCs used in LCC nodes. As noted above, the μCs have I/O that is either input or output and are under program control.

This means that the I/O lines are programmable for the application. Even in the execution of the program, the I/O pin can be changed. This capability is what allows Configuration to change without restart.

Also included in the μC is a pull-up capability. When the μC is commanded to an input on one of the GPIO pins, a large resistor is connected to the pin and the supply voltage. This ensures that the input will be "high" even when the pin is not connected. Using the sensors described earlier, the sensor output does not need a pull-up to operate correctly.

Electrically, interfacing with a μC is similar to the transistor interfacing discussed earlier. The important parameters are the levels output for high and low, and what is required to ensure input levels of high and low. Another im-

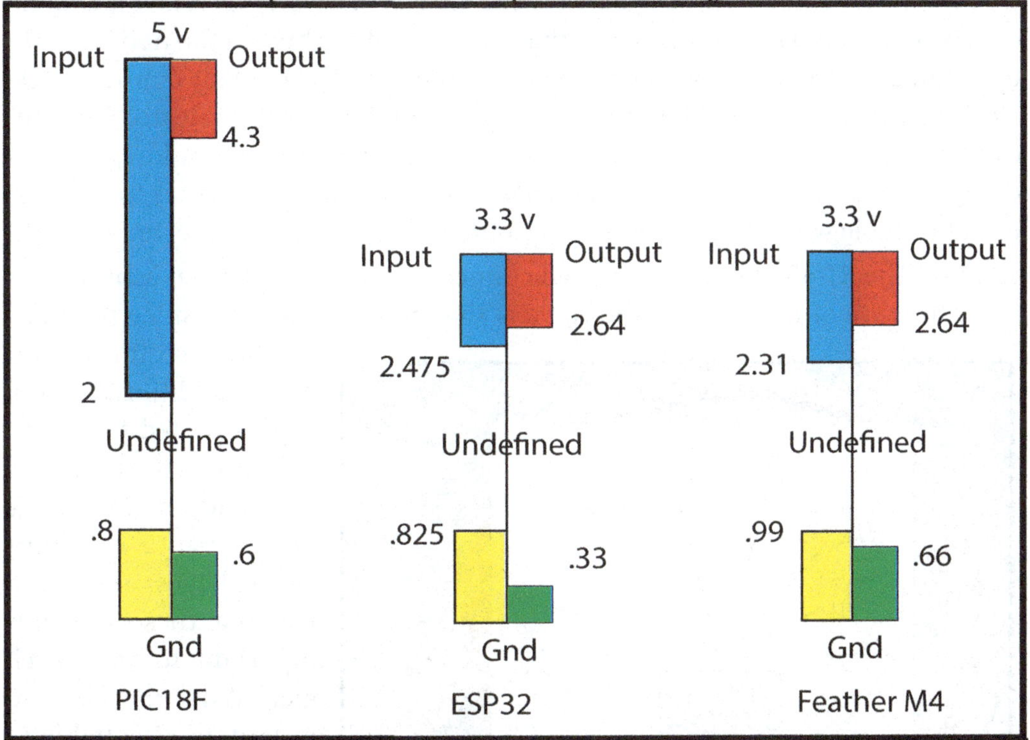


**Figure 2.11 Microcomputer Interfacing needs to stay within the Defined Limits for Reliable Operation**

portant parameter is the supply voltage which can set the other levels.

When interfacing with a μC, the input and output voltage levels must be considered. The I/O numbers shown in Figure 2.11 represent three possible μCs used in LCC. The first is typical of the TowerLCC and other RR-CirKits boards. This μC is nominally a 5-volt device which is the normal case for interfacing. Moving on, the second μC is a Arduino compatible device which is very fast and economical. The ESP32 is unique in that it has a built in WiFi capability. Finally, the Feather M4 is a very fast μC with on-board devices to provide write-able memory and transceiver interfaces for the built-in CAN.

In Figure 2.12 on page 37 the area labeled "undefined" is to be avoided. The response of the μC in this area is not guaranteed by the manufacturer. In fact, the response to a voltage in this area may change dependent on the environment and dependent on the manufacturing batch. Even if the μC performs correctly one day, the next day it may fail. If you have to replace the device, the new one might not work correctly. This kind of problem can be very frustrating, so avoid these areas.

# Details about the LCC CAN Bus

Now that we have covered the basic parts and interfaces for an LCC Node, we need to see some of the details of the CAN Bus. The CAN Bus is implemented with commercially available ICs that take care of the bits on the network. The only thing that may be of interest is the cabling and wires that connect these Nodes together. As stated before, the designers of the network selected Ethernet cables for the network. Ethernet cables are commonly available at low cost. The designation of the cable is a category, like CAT5. The requirement is for a CAT5 or better cable with 24-gauge wire. It also must have 8 wires connected to an RJ-45 plug. The RJ-45 plug is the normal connector on these Ethernet cables.

The use of the wires in the cable is shown in Figure 2.12. Since the CAN Bus only requires three wires, additional wires are designated for power and other purposes. In the standard document, the current capability and voltages are specified, but these are only important to the developer. What may be important to the hobbyist putting together a system is the power provided/used by the Node. There is a requirement to mark the maximum power used or supplied by the node on the Node. This marking can help the hobbyist to estimate power requirements. Unfortunately, the manufacturer cannot predict what may be attached to the node (LEDs, Actuators, relays, etc.) so the marking is only the bare Node power.

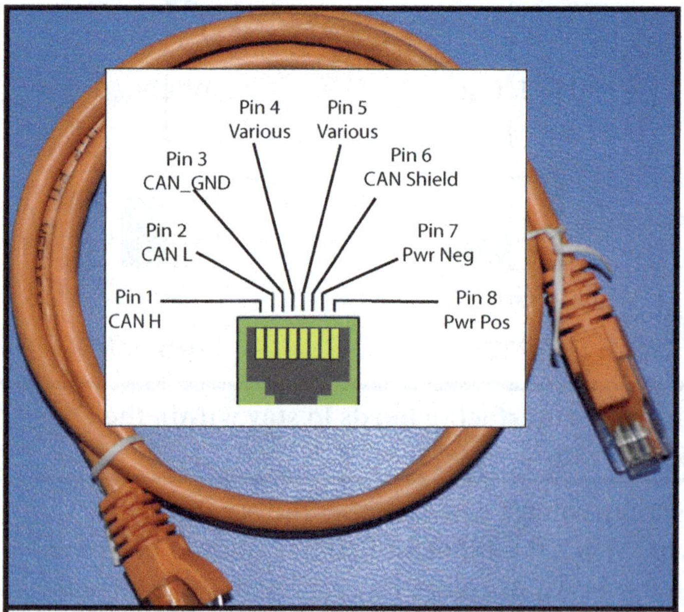


**Figure 2.12 The CAN Bus is connected with CAT5 Cables**

# Summary of Parts and I/O

This chapter is intended to provide a basis for using LCC nodes in real situations. Because we use direct current in our interfacing, Ohm's law was discussed early in this chapter. Combining Ohm's law with Kirchoff's Theorems, nearly all of our interface circuits can be analyzed.

Several simple parts used in LCC interfaces were described and shown in photos. The most common of these are simple switches and LEDs. The widespread use of LEDs makes the use of Ohm's law important for model railroad electronics.

Transistors are the most widely used device for interfacing with sensors. Several simple models of transistors are provided to aid in interfacing sensors with LCC devices. Then μC interfacing is discussed because all LCC nodes are μCs.

Future chapters will rely on some of these basics . LCC is only useful when it is interfaced with devices that are used by model railroads. More complex devices to interface with sensors and actuators (like turnout machines) are available, but the simple parts like switches and LEDs can be integrated with the simple rules in this chapter.

# Chapter 3 – A First LCC Network

To prepare for the next chapters, let us put together a minimal LCC network. As we explore this minimal LCC network, we will learn about many of the features of LCC. To see what is going on in this network, we will use the Java Model Railroad Interface (JMRI™) program. JMRI is a Open-Source program created and extended by contributors from around the world. The features of JMRI are very extensive and are well beyond the scope of this book. The subset that will be shown is oriented to LCC. JMRI becomes a Node on the LCC network and facilitates configuration and firmware update, as well as acting as a Producer/Consumer of Events. In addition, there is a viewer that allows you to see Events as they are passed on the CAN Bus. This is a very useful debugging tool.

The pieces needed to make a minimal LCC network are:

1. A computer to run JMRI — the computer brand is mostly immaterial, since JMRI runs on almost any currently available computer system, including Windows, Mac, and Linux. As noted above, JMRI will provide a Node on the network. To create a working LCC network, at least two nodes are needed.
2. An interface to the computer — the most common interface uses the computers USB port . A buffer interface device connects to the USB port and translates CAN messages to/from the computer.

This provides the hardware interface between JMRI and the LCC network.

3. An LCC Node, like a Tower-LCC+Q or Signal-LCC — A network needs at least two Nodes to be a network. This Node will be connected to I/O hardware like that described in Chapter 2.
4. Power for the network — If the Node does not supply power (voltage) to the network, then a separate power source is necessary.

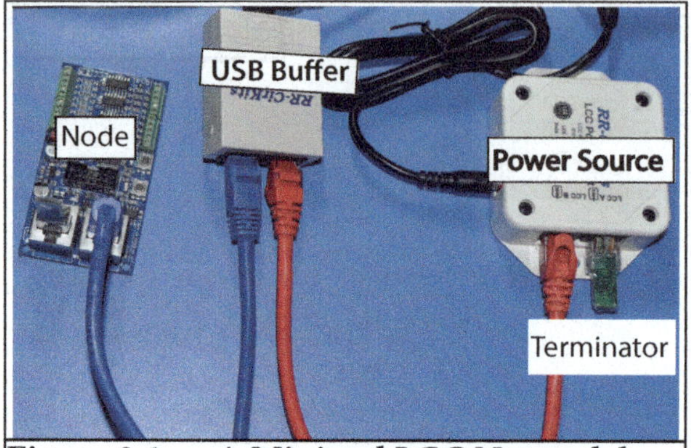


**Figure 3.1 A Minimal LCC Network has just two Nodes. (JMRI is one Node)**

5. Cables to connect between the devices — one USB cable for the USB Buffer, 1 or 2v CAT5 cables for the Nodes.
6. Two terminators — these need to be inserted into the ends of the network.

The network can be plugged together as shown in Figure 3.1 This network uses RR-CirKits devices. RR-CirKits sells a package called a starter kit which contains one USB Buffer, one Power Source (called a PowerPoint, two terminators and cables. With the addition of a Tower-LCC+Q or Signal-LCC this will put together the network shown in Figure 3.1

**Figure 3.2 This I/O Test Board provides 8 LEDs and 8 Pushbutton Switches**

Another convenient little board is the I/O Test Board shown in Figure 3.2. This little board provides 8 I/Os which can be a mixture of LEDs and pushbuttons. This board provides I/O for the new-comer. Testing configuration and learning how to set up an LCC network is much easier using this board. As shown in Figure 3.1, an LCC network can

### *Node I/O connectors*

*RR-CirKits uses a small ten-pin connector for I/O connections. These plugs, mounted on the PC board, require a ten-pin receptacle to connect to simple devices like LEDs. These receptacles are supplied with the Nodes, but the cable they connect to is not. It is a ten-conductor ribbon cable, also available from RR-CirKits and other sources including Amazon.. To reliably install the receptacles, get an IDC crimper! Working with these connectors and cables will be explained in the next chapter.*

spread out all over your computer desk without adding the prototyping boards or loose switches and LEDs. This board also allows getting started without soldering or use of other ways to put together I/Os.

To further contain the LCC sprawl, I stacked a slightly larger network together to sit on my computer desk. This stack is shown in Figure 3.3. It only differs from the network in Figure 3.1 by the addition of a TowerLCC Node. Of course, to run this network, the cables and power need to be added. Many of the screen prints that are in this book are from this stack.

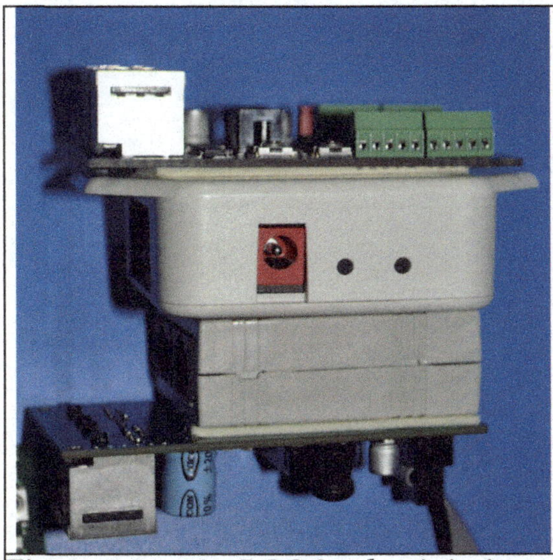

**Figure 3.3 An LCC Stack minimizes the LCC Sprawl**

It is now time to plug everything in and get power to the LCC network. If you have a recent version of your operating system, it should tell you have a new device. If not, go ahead and start up JMRI PanelPro™ anyway. The first window you will probably see will give you options to start. If one of the options is to open communications, choose that and skip the next display in Figure 3.4.

To get JMRI connected to the USB Buffer and the LCC network, choose the correct communication settings in the Preferences under Edit. As shown in Figure 3.4, click on "Edit" then click on "Preferences.

The Preferences screen should look like Figure 3.5. Set the options for the System — choose from pull-down lists. If your USB Buffer is not listed, the USB driver may be missing. Consult the manufacturer's information. Set the System

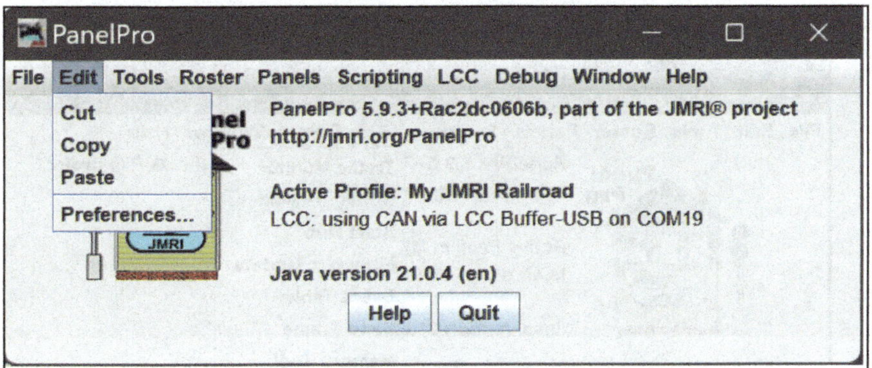


**Figure 3.4 JMRI Starts with Tabs Along the Top**

Manufacturer and System Connection as shown. The field named "Serial Port"" is another pull-down list. Often, the USB Buffer is the only entry, but if additional entrees exist, you may have to try a couple of times to find the correct one. Then click on "Save" and wait for JMRI to restart.

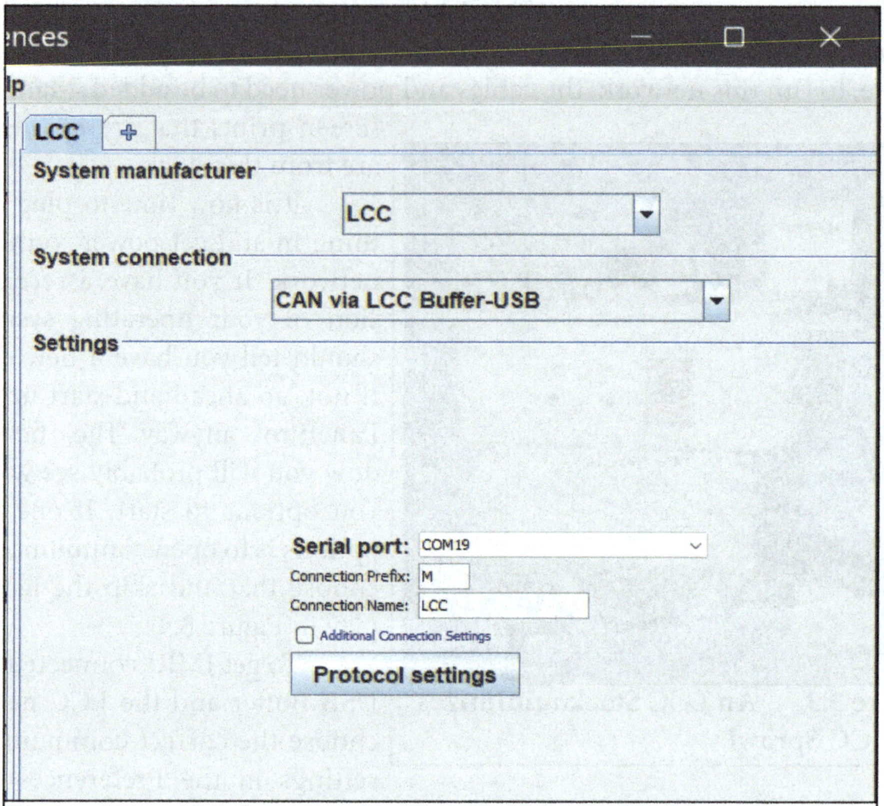


**Figure 3.5 Set Communication Options in Preferences**

Now we are ready to do something with LCC. To see the Nodes that are attached, pull down the LCC tab like Figure 3.6. For now, we will just look at the Configure Nodes option. The Traffic Monitor option allows you to watch the CAN messages on the CAN Bus. This may be useful in debugging. The Firm-

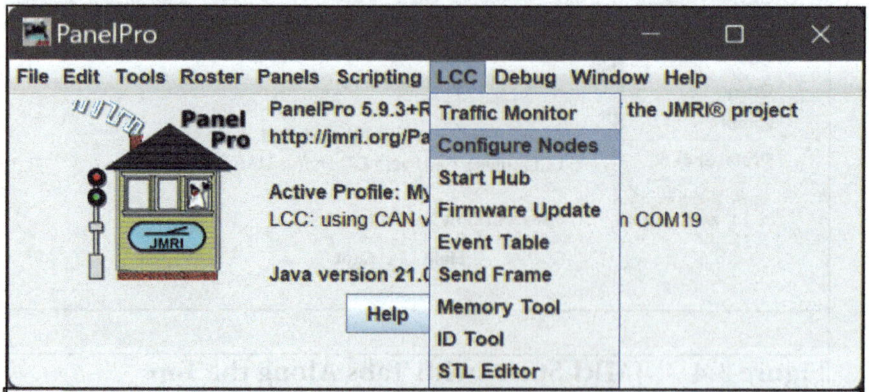


**Figure 3.6 Pull-down the LCC Tab to Enter Configure Nodes**

ware Update Option is used to keep the firmware up-to-date. We will investigate these functions later.

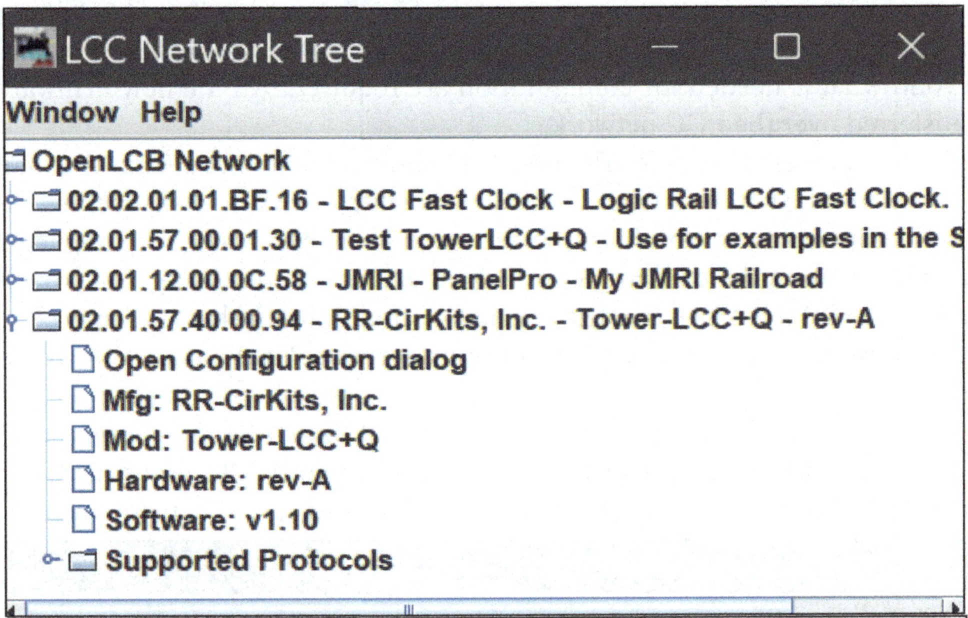


**Figure 3.7 JMRI Lists all the Nodes Currently Attached for Selection and Expansion.**

When the Configuration Tool opens, it identifies all the Nodes present in the network. In Figure 3.7, there are three nodes present. The first Node in this list is a Fast Clock like the ones discussed later in Chapter 9. Its NodeID is a Manufacturer's number, **02.02.01.11.** The Node is also identified by a description. The second node is the JMRI Node. It is given a Node ID of **02.01.12.00.0C.58** as shown in the list. The next two Nodes have a starting ID of **01.02.57** which indicates that it is from a block assigned to a manufacturer. The last three numbers are assigned by the manufacturer, and are somehow related to a model number or serial number or both. Note that the second Node has some additional information that identifies it more specifically to be a "Test TowerLCC+Q" Node. This identification is one of the items in the Configuration Data. This data is your choice. I have this Node in my LCC Stack which is a TowerLCC If you have a different Node, try to translate the terms and various choices to your Node. In Figure 3.7 the desired Node was selected (click on the tree connection for the Node ID) and a pull-down list of data is displayed. Data from the Node is displayed including manufacturer, model, hardware revision and software revision. The first item in the list is the entry to the Configuration Data. Click on the "Open configuration dialog" and JMRI will request all of the configuration data from the Node. Before we start into the Node configuration, let's make sure everything is up to date.

# Update the Firmware

One of the important functions of the LCC Network is its ability to update the Node firmware without physical access to the Node. No special connections beyond what is needed for configuration are required. All the new firmware is transferred over the LCC network.

The process is simple. The new firmware is downloaded from the manufacturer in a text format (file extension .txt)which contains hexadecimal data. You do not have to worry about the data shown on the screen, it is complete for the update. Store the data in a place where you can find it using “Save As”.Just make sure that the data is in a text format — sometimes the browser may change the extension or add to it but that is not important. Put the file where you can find it.

Start up JMRI and pull-down the LCC list like in Figure 3.6 on page 44. Click on the Firmware Update line. A window like Figure 3.8 will appear. Choose the target Id from the pull-down list, and “Select” the file. When you have done

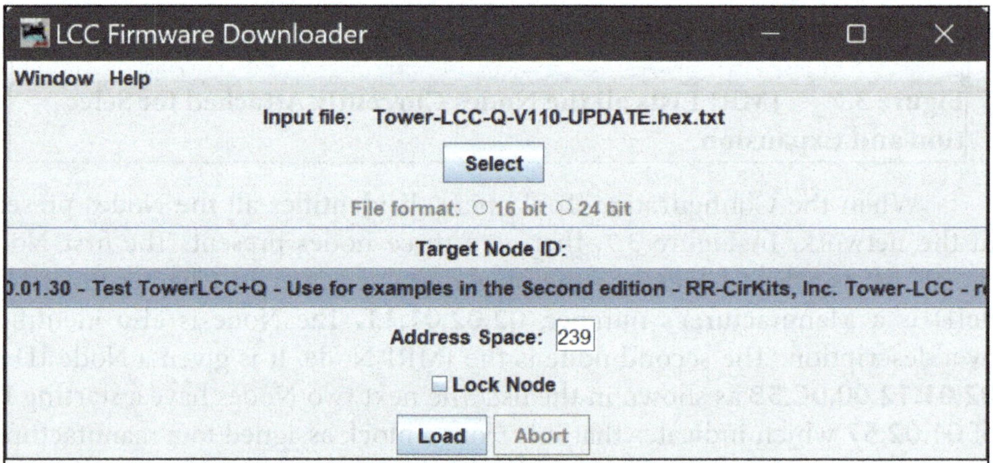


**Figure 3.8 Firmware Update is an Option in JMRI**

that, click on Load. The update will start and the progress bar on the bottom will move across to indicate completion. When the update is complete, you can exit this window and check the configuration.

Depending on the update and the manufacturer, the firmware should retain the configuration data. The CDI — which has the labels and titles — may change. Because of this, there is no guarantee that backup data stored previously will restore correctly.

# Configuration Options

The next few figures will describe the options to configure this Node. Many times the configuration action is assigning EventIDs to specific Consumers. Because more than one Node is likely to be involved, JMRI allows multiple Nodes to be open for configuration at the same time. This makes it easy to Copy and Paste EventIDs. In the next figures, note there are "Copy" and "Paste" buttons next to the EventIDs. 49

When you click on the "Open configuration dialog" line, there will be a wait while the data is being retrieved. While this is being transfered, several

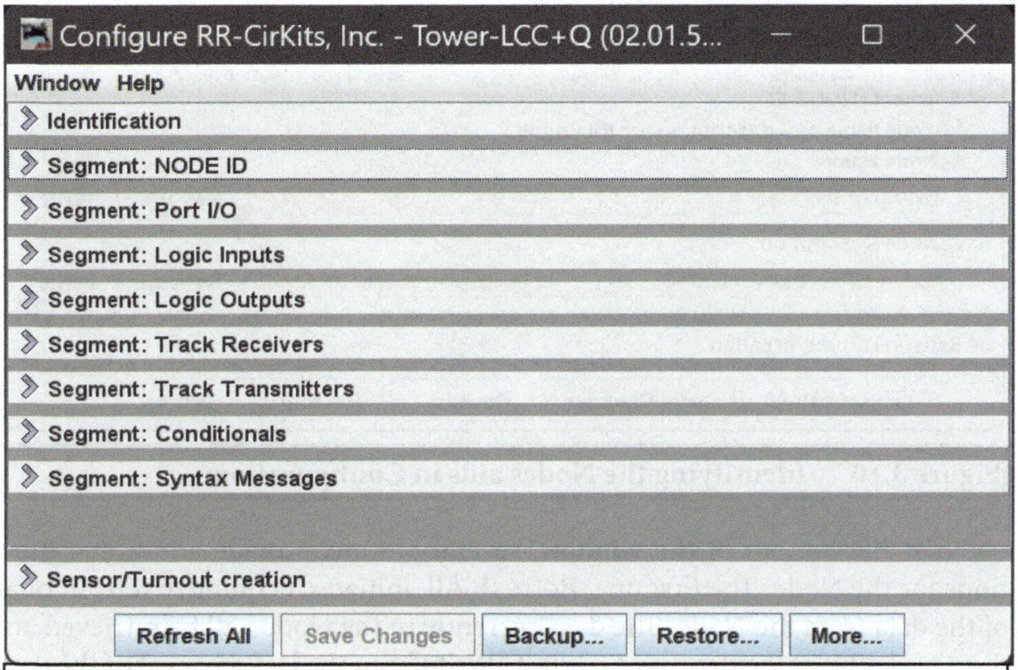


**Figure 3.9 Many Segments are listed for the TowerLCC+Q.**

lights (LEDs) on the board and the terminators will flash furiously. This is an indication of network activity. When the LEDs stop flashing, the configuration data has been retrieved.

Selecting a window is dependent on what you want to do. As a first effort, we are going to limit this to just the first three items. The rest of these are primarily used in signaling applications and will be covered in later chapters. For now, we will concentrate on the I/O applications.

Starting with the first item, it is opened by clicking on the arrow (>) on the left. This will open up the information identifying the Node. Nothing in this can be modified. When the data is retrieved, a new window will open up. Figure 3.9 shows the first "page" of data to consider. (I am using the word "page" to refer to the section of the configuration data displayed within one window. To move "down" the pages, use the scroll bar on the side or use the wheel on your mouse.) This page is data for the overall Node.

Opening the second sub-page displays the User Identification data as sown in Figure 3.11. This information is displayed on the first Node list in Figure 3.7 on page 45. Adding this information helps select the correct Node when working with multiple Nodes. Note that the changed information is highlighted in yellow. This indicates that the new data has not yet been stored in the Node. To store one of these changes, the **Write** button next to the data will store the data and remove the highlighting.

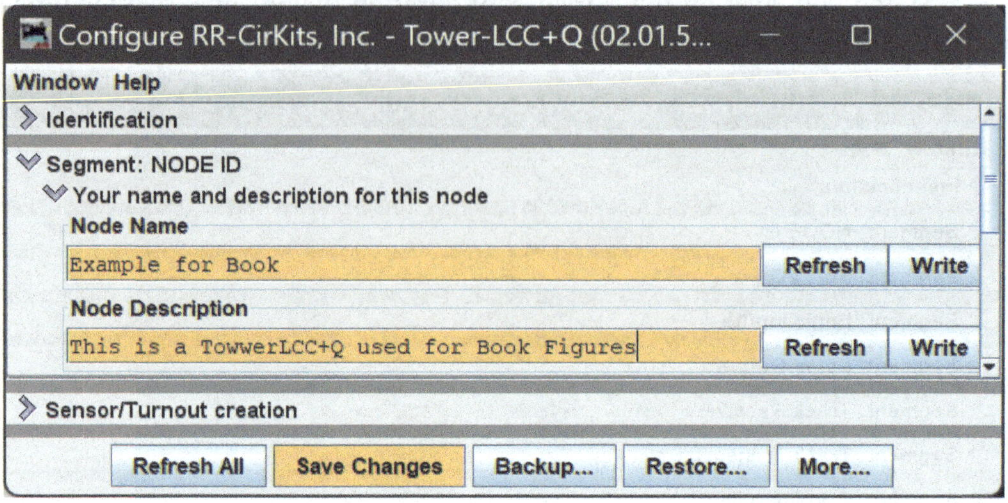


**Figure 3.10 Identifying the Nodes aids in Configuration**

At the bottom of the window, several buttons provide additional functions for the Node. The first one, **Refresh All**, initiates a complete refresh of all of the data from the Node. The data currently in the Node will be retrieved and displayed. This is the data as of the last "**Write**", not the last retrieval of the data.

The second button, **Save Changes**, is very convenient when you are making several changes in different parts of the data. **Save Changes** writes all changed data to the Node and erases the highlighting. That ensures that you will not forget to write a change.

The two buttons labeled "**Backup...**" and "**Restore...**" are used to Backup your work at a point where you are going to make changes. These functions request a location to store/retrieve the data. It is probably a good idea to backup a Node as soon as you first open it. That way, you can always return to the original configuration. However, **Backup** literally copies everything from the Node. All the **EventIDs** and choices made in Configuration will be written to a file of your choice. The **Backup** file is a plain text file, so you can read and even modify it. **Restore** writes into the Node exactly what was stored, so restoring a different Node will change all the default **EventIDs** as well as the events changed before the **Backup**.. **Restore** does not change the Node **Identification** which is applied at the factory; Node **Identification** is Globally Unique and changing it would risk having duplicate nodes.

The final choice on the bottom of this window is **More**. **More** expands to a list of new choices. These choices effect the entire Node. The first item is **Reboot**. **Reboot** does exactly that, it **Reboots** the node, forcing a Startup. **Update**

**Complete** may be similar because it notifies the Node that all changes are complete. The action that the Node takes is up to the Node implementation. Only the new SPROG nodes implement this feature.

The next choice under **More** is the **Factory Reset** option. As implemented by RR-CirKits, **Factory Reset** completely initializes the **EventIDs** and all features in the Node. The Node ID is also changed to ensure a Globally Unique ID. This action makes the Node a clean start. If you are re-purposing a Node, this is a good option to ensure that the Node is not hiding some unwanted events.

The final two features in **More** are related to JMRI features that are not covered in this book. If you study JMRI PanelPro, you will learn about what these support in JMRI.

Two user defined fields are available in this segment. The **Node Name** and **Node Description** are displayed on the Node list when the LCC Node tree is displayed. These two fields provide a reference to the use and location of the Node. In future chapters we will create documentation for each Node that will use these notations.

## Port/Line Configuration

The next segment shown in Figure 3.9 on page 47 is labeled "Port I/O." This segment is where the individual "Lines" are configured. Opening this segment displays 16 tabs to select individual "Lines." When you click on one of the tabs, a Line is selected. Think of a Line as an I/O pin. In Figure 3.11, the first line, Line 1 is selected. Each Line has a user provided **Line Description** to identify it. An entry in this field will be displayed next to the "Line #" on the tab, making it easy to identify what each Line does. Next you need to designate whether the Line is an Input, Output or both (What do you mean "Both:? This is explained later.). A pull-down list opens up when you click on the list arrow for **Output Function** or **Input Function**. Just below these function choices is a pull-down to chose the "ON" state — is "ON" High or Low. Chose High or Low based on the connected item.

What do those things mean in the **Output Function** pull-down list? The choices are **None**, **Steady**, **Pulse**, **Blink A** or **Blink B**. The terms indicate the type of output, **None** inhibits any output, **Steady** means continuous, **Pulse** means a single pulse, and **Blink** means alternating between Off and On.

**Pulse** and **Blink** need timers to set the time delays. The timers are set

> ***Node Default is Normal Input***
>
> *By default, the Nodes are shipped as* **Normal** *Inputs. Make sure you clear the Lines that are Outputs by setting the* **Line Function** *to* **None**. *It is also a good idea to clear any EventID action to None.*
>
> *If you do not disable the* **Input Function** *the node will be in the Sampling mode, which may affect the performance of some I/O boards.*

in the area just below the Output and Input data. The two timers set the On Time of the output and the Off time. These will let you determine the pulse length, the blinking period, and other things.

The **Input Function** pull-down is simply **None, Normal** and **Alternating**. None means no input and **Normal** means input is OFF or ON. **Alternating** means that it will alternate between ON and OFF with each input. This is especially useful to implement a push-on/push-off function with a simple push-button. When we look at the Events generated by this input when set to **Normal**, there will be one event action associated with the "ON" state when a button is pushed and another with the"OFF" state when the button is released . When **Alternating,** the first push will turn the input ON; a second push will turn the input Off.

> ### *A Feature of Alternating Input*
>
> *A nice feature of the Alternate version of Input is that it allows multiple push-buttons to synchronize. This means that whichever button is pressed, the Event sent is the alternate Event. If button 1 is pressed it sends Event On (or Off.) Then when either button 1 or button 2 is pressed, it will send Event Off (or On.) So whichever button is pressed, the alternate event is sent. This allows multiple control panels in different locations to change the state back and forth just the way we want it to.*

The **Delay** section of the segment shown in Figure 3.11 on page 49 defines two Interval Timers. These are used to set the time to turn ON the Line (**Interval 1**) and the time to turn it OFF (**Interval 2**). The times are set in the **Delay Time** block, and the units are in the pull-down list. Units can be **Milliseconds, Seconds**, or **Minutes.**

## Setting the Events

The next section in the Line segment is what makes LCC work. Associating Events with inputs and outputs connects the Node to the rest of the network. Moving down the segment (after the Delay Section) to the **Commands/Consumers** section we will select the actions to associate with Events. Starting with the **Commands/Consumers** at the top of Figure 3.12. This is where the events that change some state in the node are defined.

The tabs below the **Commands/Consumers** and **Actions/Producers** allow you to select an Event to associate with an output or input. Six Events can be designated for each output/input line, with different actions on output or types of input. Selecting one of these tabs selects a default EventID for use. These EventIDs will be unique to this Node since they are based on the unique NodeID. Next to the EventID is a set of buttons. The **Refresh** and **Write** buttons have been discussed earlier. **More** includes a **Trigger** action which sends that EventID to the network. Possibly the most used buttons will be the **Copy** and **Paste** buttons. Using **Copy** to make a copy of an EventID from one Line (on any Node that is in the configuration list) and then **Paste** it into the EventID field of

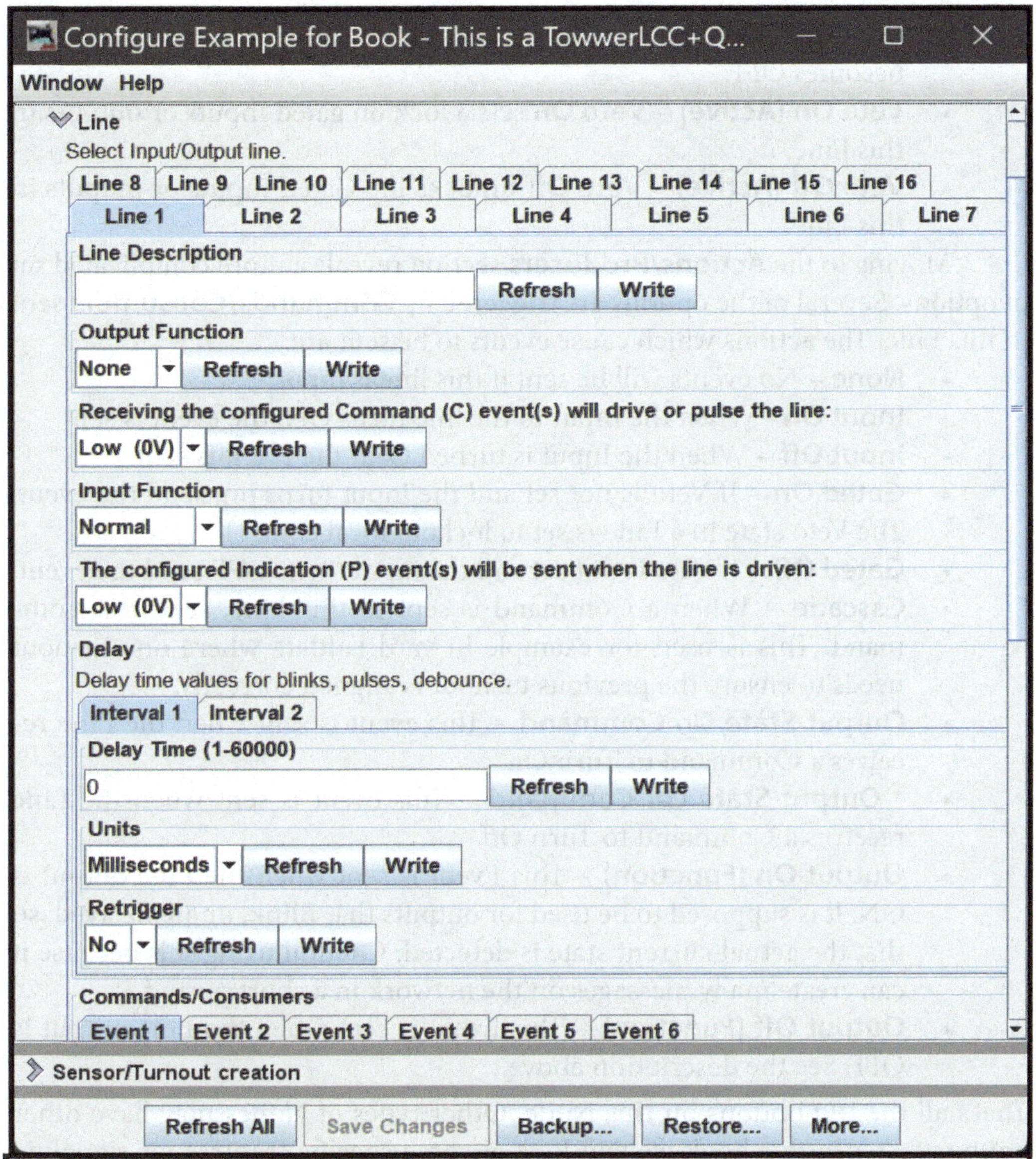


**Figure 3.11 Each Line has a Tab to set its Characteristics**

another Line saves a lot of time and reduces errors. If you see that two or three EventIDs are similar, for example, two adjacent EventIDs for a pushbutton may send ON and then OFF. The EventIDs will probably only differ in the last two digits, so **Paste** and then modify the last two digits.

Pulling down the "this Line state will be changed to" pull-down, a list of possible actions appears. The options are:

- **None** – No output state is set.
- **On (Line Active)** – The output of the Line is set to the ON state as defined by the entry shown in Figure 3.11 on page 49.
- **Off (Line Active)** – The output of the Line is set to the OFF state as defined by the entry shown in

- **Change *Toggle** – The Output is changed – OFF becomes ON or ON becomes OFF.
- **Veto On (Active)** – **Veto On** sets a lock on gated inputs or outputs in this Line.
- **Veto Off (Active)** – **Veto Off** unlocks the Gated Inputs or outputs in this Line.

Moving to the **Actions/Producers** section reveals a more complicated set of options. Several of the options are triggered by **Commands/Consumers** sent to this Line. The actions which cause events to be sent are:,

- **None** – No events will be sent if this line is input.
- **Input On** –When the input to the line turns ON, the event is sent.
- **Input Off** – When the Input is turned OFF, the event is sent.
- **Gated On** – If Veto is not set and the input turns on, send the event. The Veto state in a Line is set to lockout local control.
- **Gated Off** – If Veto is not set and the input turns OFF send the event.
- **Cascade** – When a Command is sent to this Line, send this command. This is used for example in yard ladders where one turnout needs to ensure the previous turnout is aligned correctly.
- **Output State On Command** – This event is sent when the Line receives a Command to Turn On.
- * **Output State Off Command** – This event is sent when the Line receives a Command to Turn Off
- **Output On (Function)** – This Event is sent when the Line output is ON. It is supposed to be used for outputs that Blink, or are delayed, so that the actual current state is detected. Caution using this because it can create many messages on the network in a short period.
- **Output Off (Function)** – This Event is sent when the Line output is OFF. See the description above.

That's all the I/O options on this Node. Other types of Nodes may have other options. Later in this book we will look at some specific features for signaling and clocks. Right now, let's step through configuring a Node to light up an LED when a button is pressed.

### Setup for a Simple LED button

After you choose the event action, the Producer is ready to work, because the EventID is already filled in. Line 1 shown in Figure 3.12 on page 52. Behind the Event 1 in this Figure lurks Event2. Event2 is one number larger (07 instead of 06) and is set to respond when the input is OFF. The next thing to do is to associate a Consumer action with the EventIDs. Figure 3.12 shows the EventID for Event 1. Event 2 will have the next EventID, in this case **02.01.57.00.01.30.00.07** and should be set to occur on "Input Off." These two will be used for the Consumer.

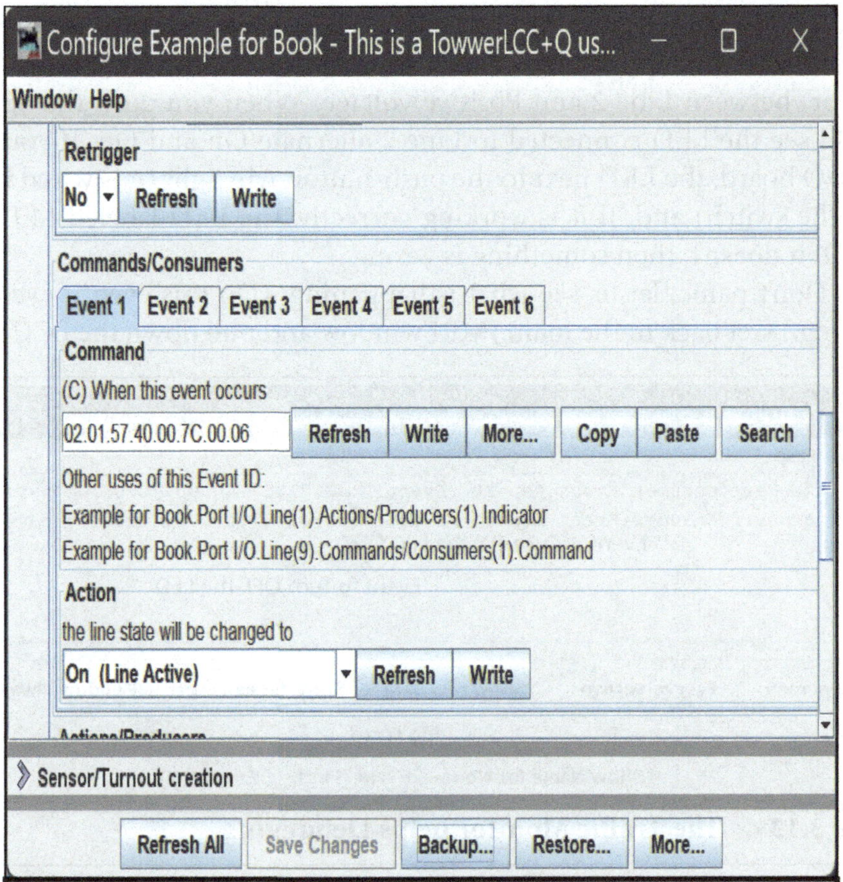


**Figure 3.12 Set up the Consumer Options for Line 2Like This to Test**

The Consumer is set-up in a similar way, with a pull-down list for the action caused by the Event identifier in the EventID field. The result for Line 2, Event 1 looks like Figure 3.12. Note that the EventID is the same as in Figure 3.12. In the Producer, the EventID was copied using the **Copy** button, then when the Consumer event was opened, the **Paste** button put the correct EventID into the field. Also note in both cases, there is a colored down arrow on the list, indicating that the field has changes. Either the **Write** button next to the field or the **Save Changes** button will save all of these changes. As before, the second Event will be set-up in a similar fashion, using the Off action in the pull-down list.

# Testing the LCC Network

If you have been following along, you may already have configured this Node to the options that I have described. If not, I'll wait while you do …Okay, now we need to test the changes we have put in. The minimum that we need to test this network is a pushbutton connected to Line 1. Also, to display the output, we need an LED connected on Line 2. If you have the Test I/O board, plug it into the connector nearest the LCC cables. Otherwise, use whatever method you prefer to connect a push-button between Line 1 and Ground and an LED (with resistor) between Line 2 and Positive voltage. When you push the button, you should see the LED connected to Line 2 alternate On and Off. If you have the Test I/O board, the LED next to the push-button will light (it is wired in parallel with the switch) and, if it is working correctly, the LED labeled LED2 should light. If it doesn't, then something is wrong.

Don't panic, let us see what is happening. (Do this even if everything is working.) Get back to the main JMRI window and pull down the LCC list. Click

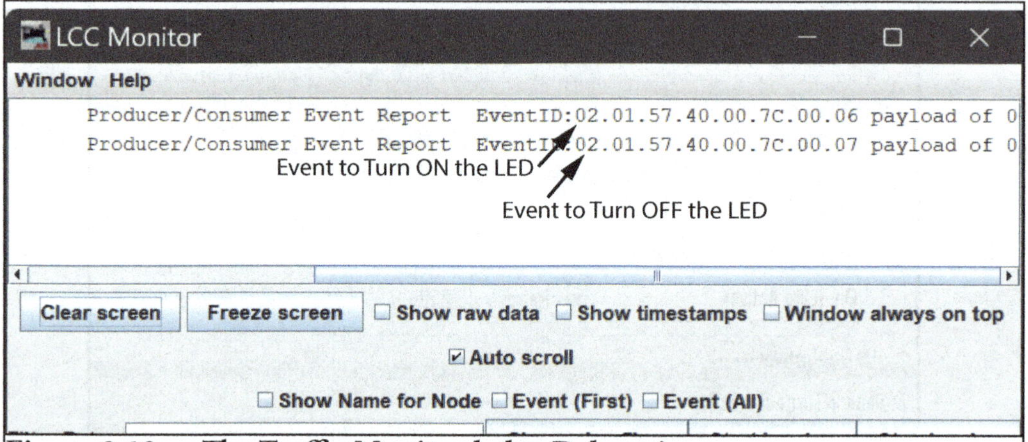


**Figure 3.13 The Traffic Monitor helps Debugging.**

on Traffic Monitor. A Window will open up to monitor the LCC network. If you press the button again, you should see line entries in the window that look something like Figure 3.13. If the EventID has a **07** at the end, press the button again. If nothing happens (no Event is shown) then the Producer did not send an Event. Recheck the Producer set-up and the push-button connection. If that is alright, try again.

If you are seeing the Event in the Traffic Monitor, then there must be something wrong on the Consumer end. Check all of the connections and options on that end, then try again. This tool helps split the testing between the Producer and the Consumer.

When pressing a button (or turning on any other sensor) causes an unexpected result, go back to this monitor to see what Events are actually being created.

In my test case, I wanted to be able to catch just the Input ON event, so I changed the **Input Function** (see Figure 3.11 on page 49) to **Alternating**. Now I can press the button once and only one event is sent. Try this on your setup and then explore all the other things we have discussed.

# Sharing a Line with Sampling

Earlier I said there is a feature where a Line is used as both a Producer and a Consumer. The feature is called Sampling, referring to the way it is implemented. The reason why this feature was developed is to save Nodes. A typical use is in the assembly of a control panel for turnout control. A pushbutton can be used for the turnout (especially if it is an eNtry/eXit (N/X)yard control panel) and an LED provides the indication that this is the current state. Combining these on one Line cuts the number of Lines in half and therefore reduces the number of Nodes.

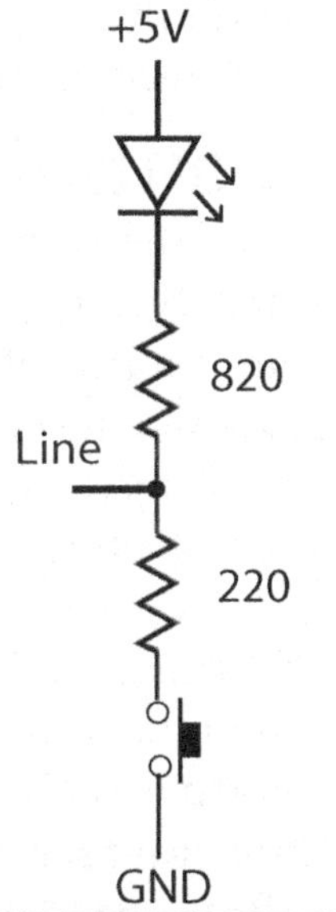


**Figure 3.14 Sampling requires resistors for Current Limiting**

Sampling is performed as it is named — at regular intervals the Line is changed from Output (controlling an LED, for example) and the Line is sampled as an Input for a short time. As noted in Chapter 2, µCs have the ability to switch Line types in their programs. This sample can therefore be very short, maybe only microseconds. These short sample times are completely invisible to many outputs like LEDs. Studies show that buttons are pressed for 10s of milliseconds, which means the sampling will probably have 100s of samples while the button is On. This means that the Sampling is a very reliable form of detecting a button press.

But wait a minute. Let's consider what we are connecting. As noted before, the Node is a µC with specific Input/Output characteristics. If the connection does not meet the levels in Figure 2.12 on page 37, we cannot be sure that the sampling will work. Another consideration is the current maximum that the Line can source or sink. Analyzing all the different requirements and different modes , I determined that the Figure 3.14 would provide adequate brightness for an indicator and still meet the µC I/O requirements.

For those that do not want to assemble their own shared lines with sampling, RR-CirKits sells a ready made button LED combination called a Button QuikLink. This is a nice packaged Illuminated Pushbutton with the resistors and wiring to use for a control panel or other use. Combined with the Breakout QuikLink board, eight QuikLinks can be plugged into a Node. All connections are standard cables — 10 pin connectors and 3 wires to the breakout board. The only real problem besides the physical size is that there are several solder pads on each button assembly that determine the color of the LED for On or Off. Additional options include Alternating and dim or bright.

Sampling depends on the output device that is being controlled, in the above case, and a LED is the output device. Another possibility would be the SMD8 to control a Tortoise turnout motor. In this case, the SMD8 would replace the LED and resistor in Figure 3.14 on page 55. The switch resistor can be increased in value, because the input impedance is over 5K ohms. A 1000 ohm resistor is recommended.

# Controlling I/O with Veto and Gated

In the **Commands/Consumers** and the **Action/Producers** sections, there are several options which are proceeded with the word **Gated**. These options are controlled by the state of the **Veto** variable associated with the Line. If **Veto** is set, the Gated **Consumer** command will not be executed, even if that EventID is received. Other EventIDs can cause the Output to change, so it can still control a turnout, for example. On the **Producer** events, Gated Inputs will prevent any event coming from the input on that Line. This prevents fascia control of turnouts and allows a dispatcher to control the turnouts without interference from the operators or visitors.

**Veto** is set by an EventID received by the Line **Consumer**. Even though the **Consumer** seems to be part of **Output**, it can be set without an **Output** setting. The **Veto ON** and **Veto OFF** commands provide the control for this feature. These Events would only be created by the Dispatcher. After sending a **Veto ON**, the dispatcher would then control individual turnouts with non-Gated events, until the Operating session is over. Then to clear the Vetoes, a **Veto OFF** event would be sent to all turnouts.

# Cascade to Ripple Down a Ladder

One of the Producer options is to create a **Cascade** Event. This feature is

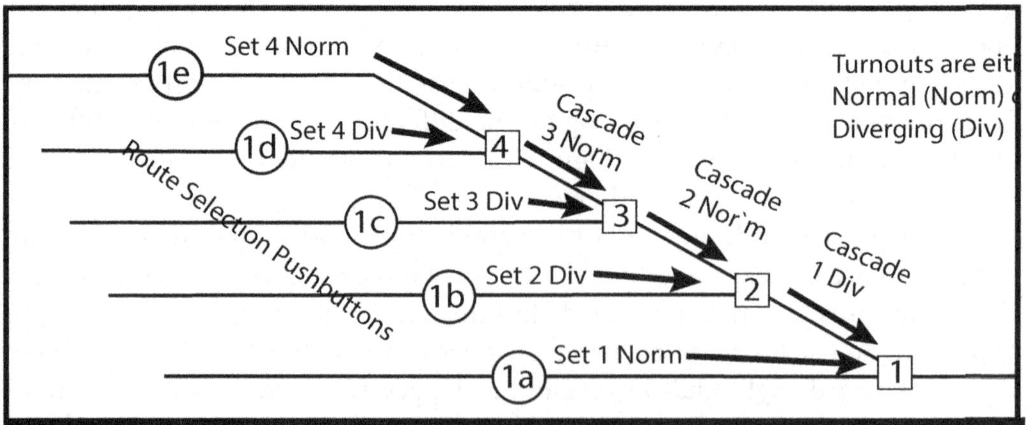


**Figure 3.15 Cascade makes Yard Ladders Easy.**

especially useful for eNtry/eXiy yard ladders. The technique is shown in Figure

3.15. Each button sets one turnout. Each turnout down from the one on the desired track is set by the previous turnout's **Cascade** Event. Since any change of an upstream turnout requires the downstream turnout set in a direction, **Cascade** works whichever way the turnout is commanded, even if the turnout doesn't move.. All turnouts except 1 need to set the downstream turnout so that trains can reach that turnout.

**Output State ON Command** and **Output State Off Command** are also cascade type, but only if one direction is commanded. These options can be used to provide a new event when the output is commanded to a specific state by more than one event.

# Events when an Output is On or Off

The final two options in the **Producer** Events cause Events when the Output changes state. These options are intended to be used when the Output is **Pulse**, **Blink A** or **Blink B**. If, for example, the Output state for a Blink Output could generate a strobing effect on another output which is the negative of the original Output. Using these two Outputs for the lights on a crossing signal. Flashing alternately between the two would be the result.

A possibility for the **Pulse** would be a control for some animation. A button push would start the **Pulse** and the animation. If the pulse runs out, the animation would turn off. If the **Pulse** is **Retriggered**, the pulse would be lengthened. If at the end of the animation, the animation needs to be "parked," an event generated from this option will trigger the motion to "park" everything and be ready for the next move.

Overuse of these options can produce many events on the network, Overloading the network reduces the response, so be cautious with these options.

# An Alternate Configuration Tool

The configuration and monitoring in the preceding sections has exclusively used the JMRI Tool. Deepwoods Software offers an alternative tool. Unfortunately, that tool only runs on Linux or Mac-OS computers. Since over 90% of all computers (excluding tablets and smart phones) use some variety of Windows, I will continue to use JMRI as my examples. Since most of the labels and field definitions are contained in the Node's memory, the transition to a different tool should be simple.

# SPROG™ LCC Boards

Recently a new manufacturer of LCC boards has delivered some LCC Nodes. The company, SPROG, is offering two nearly identical boards and support boards. These boards are primarily designed to control Servo motors, but can be applied for other uses. Four of the boards are shown in Figure 3.16 on page 58. These four boards are sufficient to test the SPROG LCC. The two at the left are the support boards for powering the network (PWR-LCC) and connecting to JMRI (USB-LCC). The active LCC Nodes are the IO-LCC and SERVO-LCC boards. These two boards are effectively the same except for the interface components on the SERVO-LCC. The SERVO-LCC board has connectors and resistors or drivers to control Servo Motors or other devices. Other than that, the programming seems to be identical.

First impressions include:

- The SPROG boards are larger than the RR-CitKits boards. They are different sizes.

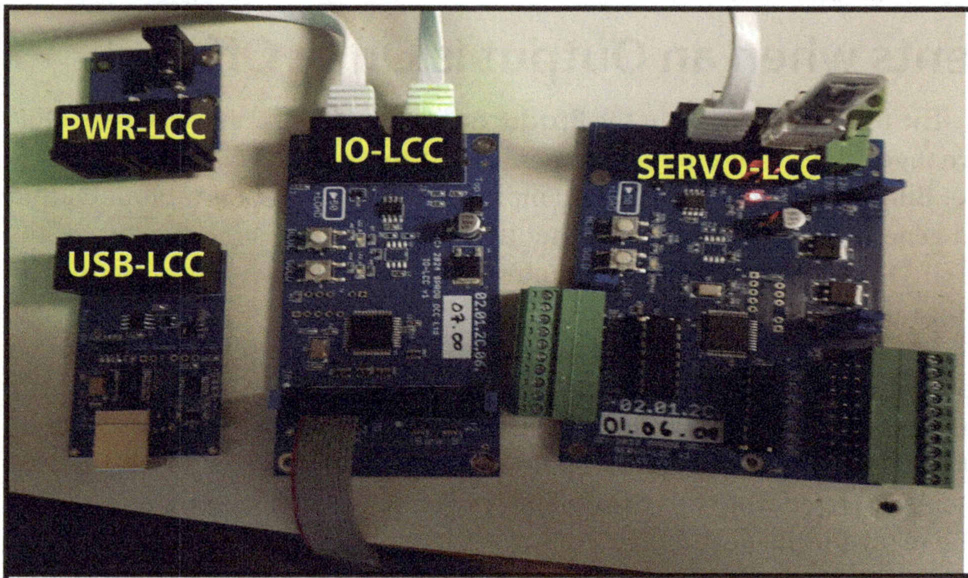


**Figure 3.16 SPROG™ has developed a New Line of LCC Products**

- RJ46 connectors are parallel with the board. This takes up a little more horizontal space.
- I/O connectors are different – the IO-LCC has 10 pin connectors like RR-Cirkits (but watch out for the numbering) and the SERVO-LCC has European style plug-in screw connectors. Matching connectors are provided.
- The 10 pin connectors on the IO-LCC are numbered like the connectors, as opposed to RR-CirKits numbers its Lines backwards. That connects pin1 on the RR-CitKits Connector to Line 8. Fortunately, pins 5 and 6 are o volts and 5 volts on both SPROG and RR-CirKits.
- As required by the Standards, when plugged into my network with 15 volt power, it immediately operated.
- There are LEDs on the RJ45 connectors that display Power On, CAN Transmit and CAN Receive.

- The Node ID is printed on the top of the board where it is easy to see after installation.
- The ButtonLED Test Board provided by SPROG is a nice linear layout, buy it needs wiring to connect to either LCC board. Cut and strip 10 wires to connect between two plugs, which uses up two plugs and now you can't use one for I/O.
- There are several jumper choices on the SERVO-LCC that configure power options. Default positions work with the CAN network power. For details, see the User Manual.
- An option mentioned in the User Manual increases the output current capability by replacing two ICs with driver ICs. The driver ICs allow up too 400 ma. total for the IC. The IC is a TBD62083 type.

Digging a little deeper, I started up JMRI to check the Configuration Data. As expected, both boards showed up in the configuration list. Moving on, the SE-VO-LCC board configuration was opened to find the window shown in Figure 3.17 on page 59. Many of these are similar to previous selections. I cheated, because I have already added the Node Description, which is the same as the previous Node ID.

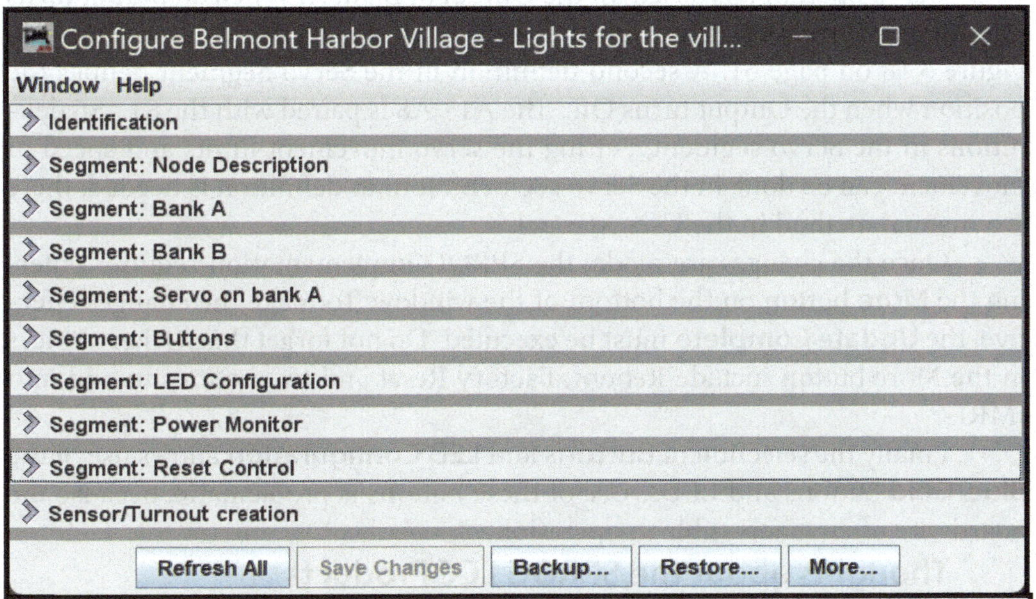


**Figure 3.17 SPROG SERVO-LCC Configuration segments have different names for Similar Functions**

The standard buttons show up along the bottom, but the **More** button will need some additional explanation. Looking at the choices, one thing is clear – there is no Logic or Conditionals to support signaling. The other segment choice seems to be for Servo control. The Reset Control option is an interesting option.

Expanding it gives you a set of choices to apply to the next **Reset**. The one which can be used here is to reset to **Factory Defaults.** Unlike RR-CirKits, this actually sets all the choices and events as they were originally. When I am playing with a new board, I like this option, but be cautious, there might be some hidden events on other boards that used the original EventIDs. Expanding a window for the Bank A set of channels (Lines in the RR-CirKits Configuration) produces the window shown in Figure 3.18. The first selection is to decide if this entire bank is **Input/Output** or **Servo**. Then the pin is selected by A1 - A8. The function of that pin is Input, Output, Sampled or some form of pulse. Sampled is the same as the RR-Cirkits, but you have to explicitly set Sampled. The times for the pulse and repeating are set next.

The next sections define Consumers and Producers. Note that the distinction is not Output or Input; The choices are whether this consumes an event or produces an event. Therefore, the Consumers are primarily setting an Output pin and the Producers set events. If you want an Input to send an event, look under Producers; If you want an event because an output happened, look under Producers and so on. Some functions are not implemented, notably the **Gated** functions and the ability to delay an Output.

Servo operation is based on the Output of Bank A. When an output turns On, the servo will move to a position defined in the Servo Segment shown in Figure 3.18 on page 61. A second definition in the Servo Segment defines the position when the Output turns Off. The A1 - A8 is paired with the S1 - S6 definitions in the Servo segment, Setting the servo movement limits and speed of movement can be done in the Servo segment. If finer definition is needed, there is a manual method in the User Manual.

Once the changes are made, the SPROG implementation requires entering the **More** button on the bottom of the window. To make the changes effective, the **Update Complete** must be executed. Do not forget this. Other choices in the More button include Reboot, Factory Reset and two actions to add into JMRI.

Finally the selection of **Buttons** and **LED Configuration** allows use of the **Blue/Gold** buttons and LEDs. Use of these buttons is problematic, because the boards are often inaccessible in operation.

## Thoughts about the SPROG LCC Products

SPROG has a good start in the LCC product line. The complete set of modules can be used to create a basic LCC network. SPROG has chosen to fill a niche that will be well received. Creating an LCC interface to Servo Motors takes care of the many cases where servos are used for turnouts. The ServoIO-LCC module seems to operate servo motors only over a limited range of angles. The Configuration is limited to 1 msec to 2 msec. This is only 90° travel.

All their modules seem to be compliant with the LCC Standards as much as I can test. The board layouts seem very good with connectors accessible and

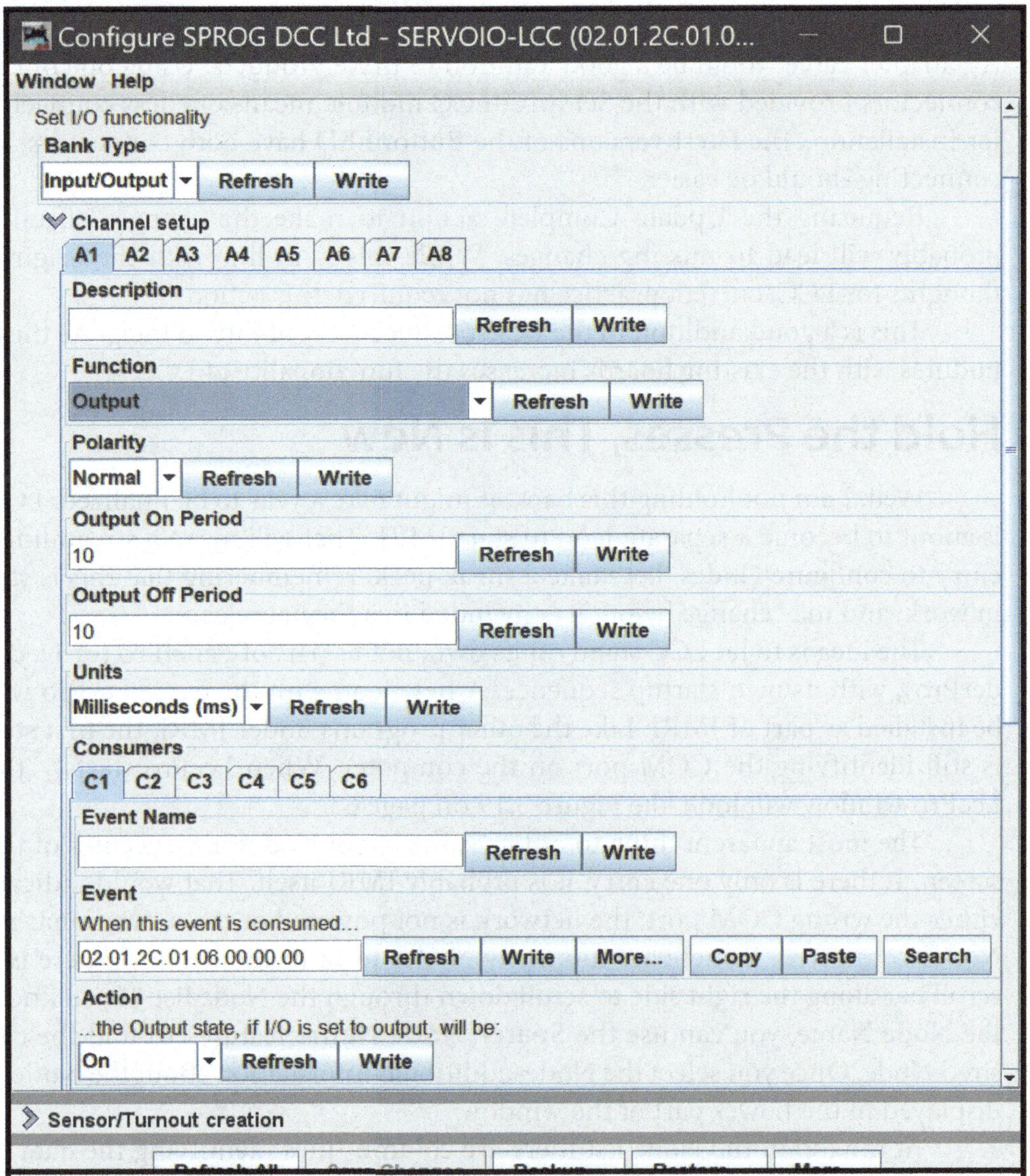


**Figure 3.18 Bank A is the Servo Control or I/O Bank. Bank B is just I/O.**

identification clearly visible. The placement of the Node ID makes it convenient to reference when mounted.

A note about power for the ServoIO-LCC. The options provided with jumpers are good, understanding is needed. When the AUX power is selected and a external supply provided, this does not power the main LCC functions. The AUX power is for the IO circuitry, especially for Servo Motors. In many ways this is good.

Some of the deficiencies include the lack of Gated inputs, and Logic functions. Both of these are special cases and not used in all installations. Another issue is the connectors on the ServoIO-LCC and the ButtonLED Test board. An adapter cable would be a good addition to these products. Using one of the connectors provided with the ServoIO-LCC module means one less connector for installation. The latest versions of the ButtonLED have both connectors, so connecting should be easier.

Requiring the Update Complete action to make the changes effective probably will lead to missing changes. While this is in line with the original thoughts for LCC, current practice has not required this action.

This is a good addition to the LCC products. The ability to integrate these nodules with the existing boards increases the functionality of LCC.

# Hold the Presses, This is New

Well, I am not holding this back, it might take a year to be finalized. LCC is about to become a separate Icon to start JMRI. That will mean a streamlined entry to configure Nodes. Let's take a sneak peek, remembering that this is still in work, and may change before it is included in a formal release.

The idea is to let LCC stand on its own, not as part of PanelPro (or DecoderPro), with its own startup sequence. A new Icon with the name LccPro will be installed as part of JMRI. Like the other programs under JMRI, the first step is still identifying the COM port on the computer. When LccPro starts, the LccPro window will look like Figure 3.19 on page 63.

The most apparent thing to notice is the list of Nodes in the center of the screen. If there is only one entry, it is probably JMRI itself. That would indicate either the wrong COM port, the network is not powered or the network has no Nodes connected. If there are too many Nodes to fit in the window, there is a scroll bar along the right side to scroll down through the Node list. If you know the Node Name, you can use the **Search Node Name** feature .To find the desired Node. Once you select the Node, additional information about the Node is displayed in the Lower part of the window.

At the top of the Node list there are column titles identifying the data in each column. Selecting (Clicking) one of these column titles will sort the data in ascending or descending order. If you Right Click on a column title, a selection list is presented to select visibility of the columns. In this example, I have hidden the Manufacturer column., There is an important column that shows the Software Version in the Node. This column lets you quickly identify the out-of-date Nodes that need to be updated,

The last two columns contain "buttons" that start features that we have already discussed. **Configuration** and **Upgrade** are the same features that we have already discussed. The Tabs and Pull down lists at the top of the window are simply entry points into the respective functions already discussed.

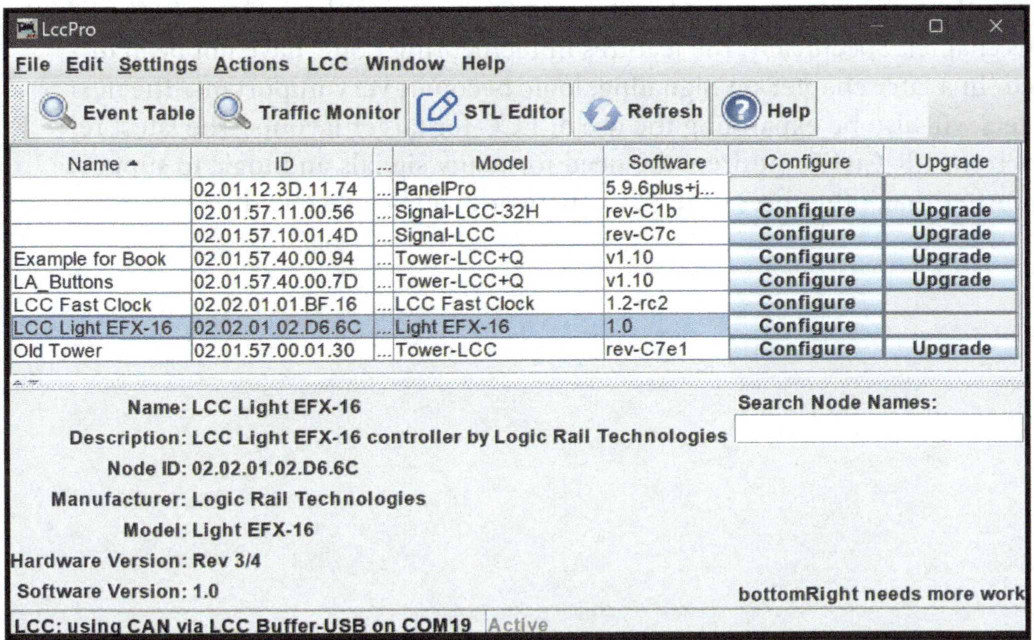


**Figure 3.19 The LccPro Window Lists the Nodes and Sets you up for Configuration.**

When this new feature is available, it will provide a streamlined path to the LCC Node manipulation. As LCC networks grow, this capability will make it easier to find the desired Nodes and provide quick access to **Configuration**.

## Summary

Congratulations, you have assembled your first LCC network! It doesn't do too much right now, but the potential is there. Just this much could be used to light an LED on a control panel to indicate that a turnout has changed position. In that case, the same Event could be used to actually throw the turnout.

In this chapter we have assembled the minimal LCC network which consists of one Node Board, one USB Buffer, a source of power, and a JMRI Node running on a computer. With this small bit of hardware, we have demonstrated the Configuration process and demonstrated the network operation. We also looked at the Traffic Monitor to help debug the configuration.

All of the boards used are from RR-CirKits. New manufacturers are starting to provide LCC products. As these come out, some small changes are likely, but the basics will remain the same. At this time, all the configuration tools operate in a similar way, and the Nodes will all use the same interface. While Node features may change, the basic Configuration process probably will not. What may change is the order of Configuration and the names of features. For example, Lines were once called Buses. Always keep the Nodes updated, at least until they are working completely

In the following chapters I will expand on some of the options discussed in this chapter. Specifically, the features implementing Logic have not been discussed. In a later chapter on Signaling, logic becomes very important. The next chapters will also be expanding the use of LCC for larger layouts. The latest releases from RR-CirKits address the need for many signals and logic to support them. After that, other uses of the LCC network will be presented.

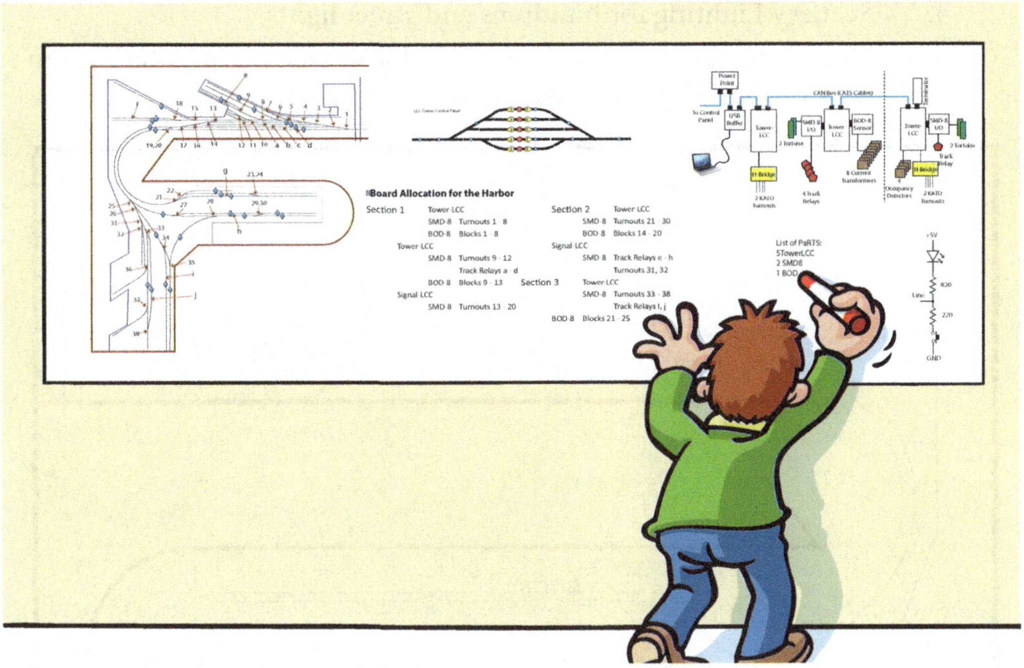


# Chapter 4 – Planning LCC

Beginning an LCC design seems to be a massive effort. Where do you start? There are so many things to consider, so many pieces to come together. Let's take it one step at a time. Paraphrasing E.B. White's recommendation, when you have more than one thing to do, do one and then the next, don't try to do them all together. In this chapter I intend to step through the planning of an LCC network and show some analysis tools that may help.

Since abstract steps are difficult to follow, I have chosen to use an example from the Belmont Shore Model Railroad. One of the areas on this layout is a harbor which represents a small portion of the Los Angeles harbor in about 1950. It is a complex area with many turnouts and track segments. It includes a wye, a small yard, and many stub tracks. It is an area roughly 16 ft. square with a aisle in the center. A drawing of the harbor is shown in Figure 4.1. As an added complication, the bottom part of the drawing can only be accessed by putting a platform up over the stairs that are under it. This means that the wiring and controls need to be well designed to limit rework.

Some of the potential functions to implement are:

1. Route switching to make it easy to select yard tracks;
2. Track power switching to allow locomotives to be parked in the yard or other locations;
3. Dual control panels including signal masts to limit walking around the peninsula while controlling a train;

4. Scenery Lighting for buildings and street lights;
5. Occupancy detection and signaling for tracks that may be hard to see over buildings and around other operators;

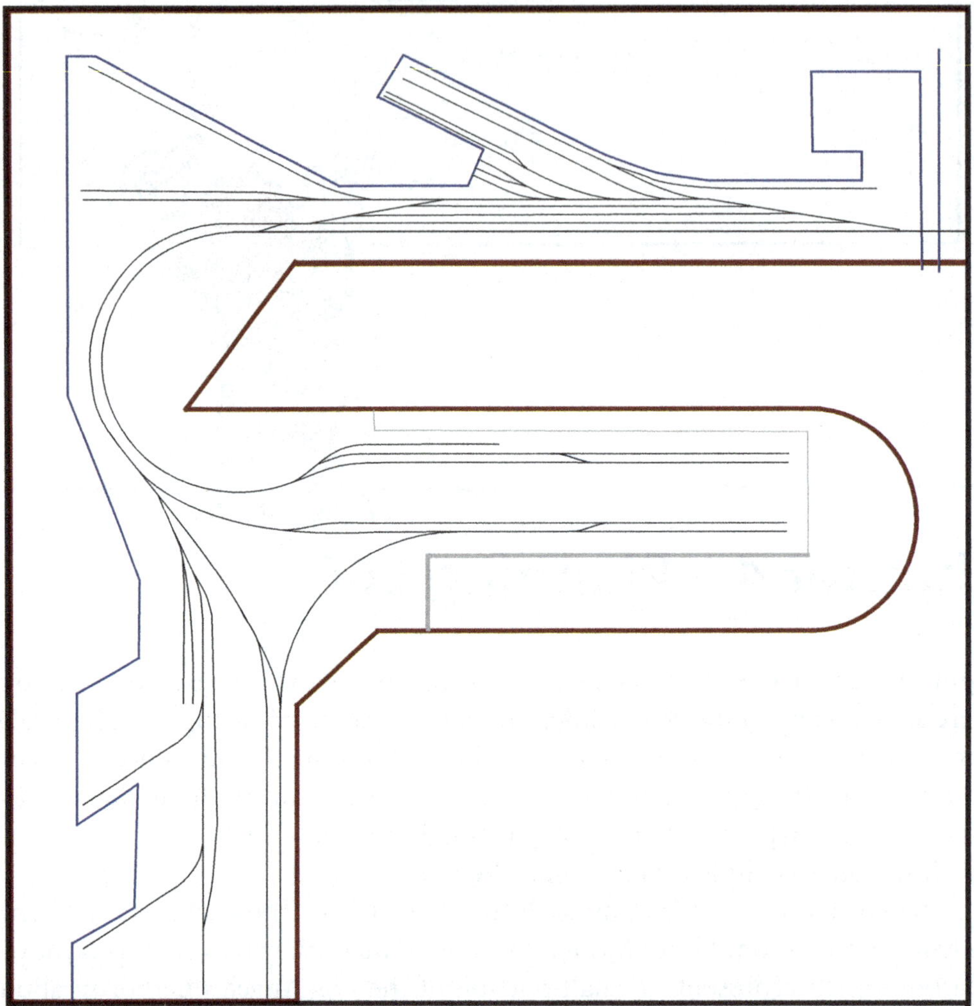

**Figure 4.1 The Belmont Shore Line Harbor has Interesting Problems to Solve with LCC**

6. Road crossing signals in a couple of places

Fortunately, LCC can be added in stages, so all of these features do not need to be installed at once. Before any LCC is installed, the basic track-work and wiring should be completed. This includes the track power, power districts (for reversing and short circuit protection) and turnout machines. Local testing of these functions should be performed so that the addition of LCC does not include fixing issues with the non-LCC pieces. For the first level of planning I will concentrate on the Operating functions of the harbor. I will leave signaling to a later chapter and will also address lighting control in future chapters.

# What do we Need?

To determine the pieces of LCC to acquire, we need to count the specific items on the layout. The important items are the turnouts, track blocks, and route switched yard tracks. Figure 4.2 shows the count of the turnouts in this area. 38 turnouts need control to switch all the tracks. Some combination of turnout machines might be possible on the crossovers, because both turnouts

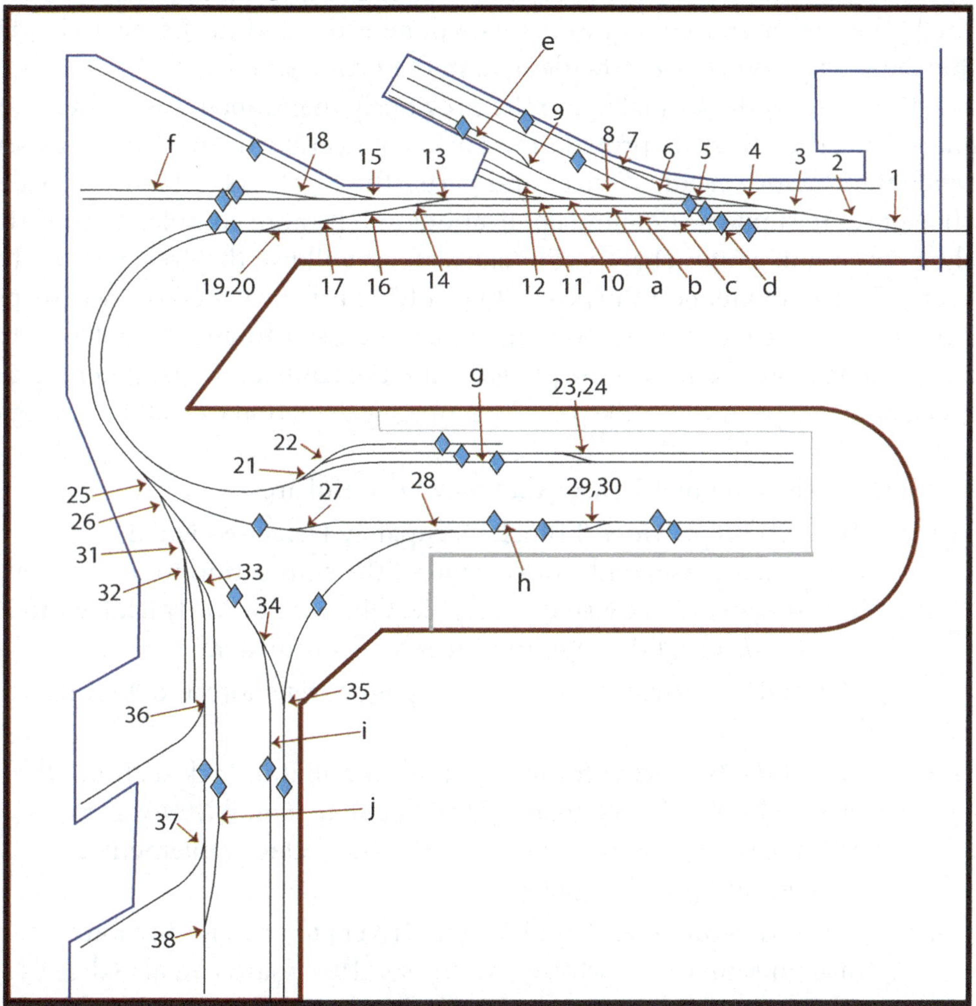


**Figure 4.2 Counting the Turnouts shows the Magnitude of the Problem**

can be thrown at the same time. Either a mechanical linkage or an electrical combination would reduce the LCC I/O required.

Counting other items, I see about 10 tracks that could be switched off to park locomotives. These locations are labeled with lower-case letters in Figure 4.2. Relays to switch the track power will be used on these tracks. Occupancy detectors are designated with diamonds on the tracks. I see about 23 diamonds to install occupancy detectors on track feeds. In raw numbers, this section requires 48 outputs and 23 inputs for the track controls.

# Available Boards for LCC Nodes

To connect to the LCC Network, the specific boards must be considered. For this example, the RR-CirKits™ boards will be used. In addition to the Nodes, support devices like a USB Buffer, power sources, and terminators must be included. The power source requirements will be estimated in the next chapter. Other support pieces will also be defined in the next chapter.

The basic Nodes available are the previously mentioned Tower-LCC and Signal-LCC. These boards provide the basic connection to the CAN Bus and provide the I/O for specific devices and parts. The Signal-LCC board is loaded with features to support signaling applications and is more complex to use. This will be covered in a later chapter. For starters, we will use the Tower-LCC. The Tower-LCC has 16 General Purpose I/O (GPIO) ports provided on two 10-pin connectors. Either the mating connectors can be used to attach cables to the board, or a daughter board can be plugged into the connector. Daughter boards are available for specific I/O, like block occupancy detection or stall motor drivers.

Some of the daughter boards that may be useful are:

1. The BOD-8 provides 8 Block Occupancy Detectors when connected to Current Transformers. The connection to the Current Transformers is a twisted pair cable. One wire of the track feed to a block is inserted through the Current Transformer.
2. The BOD-4 provides 4 Block Occupancy Detectors and 4 GPIO lines.
3. The BOD-4CP provides the interface for an interlock section. It provides 4 Block Occupancy Detectors and 4 Stall Motor Drivers. This board on a Signal-LCC provides the I/O to implement a passing siding with signals.
4. The SMD8 provides 8 Stall Motor Drivers to power 8 turnout machines like the Tortoise™ machines. This board can also drive a relay when wired correctly. It is limited to 100 ma. per output and 200 ma. for the board.
5. The SCSD8 provides 8 drivers for Single Coil Solenoid turnout machines. This board is for a single-ended solenoid.

Combinations of these daughter boards are used to assemble the I/O necessary for any set of functions. For example, using the daughter boards, 38 turnouts will require 5 SMD8 daughter boards. That is not the way to determine the boards needed, because the boards need to be located in logical sections. So, the next step is to split the problem by sections.

# Allocating Boards

The first section of the harbor is the top third of the original drawing. This is also the entry into the harbor from the Belmont Yard. Figure 4.3 shows the functions in this section of track. Counting the functions to interface, I see 20 turnouts, 3 track power relays and 13 block detectors. Allocating boards to these functions, I could use 3 SMD8 daughter boards to drive the turnout motors and the relays, and 2 BOD-8 for the block detectors. Since that will require 3 nodes (each Tower-LCC can support 2 daughter boards) one Node will be only half used. I might at this time consider that I will be adding signaling, so maybe I will use 2 Tower-LCC and 1 Signal-LCC node.

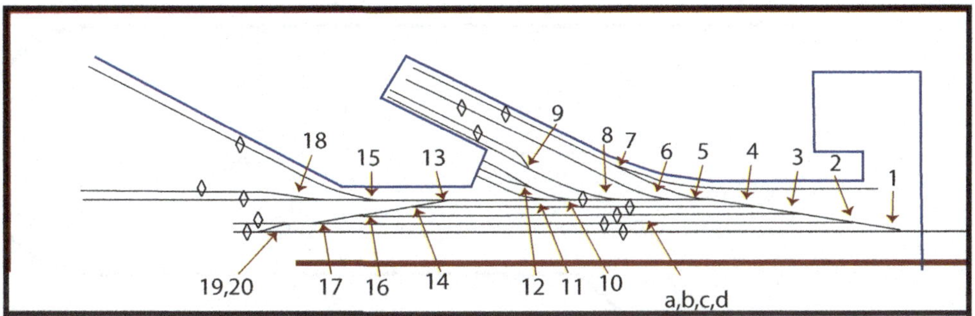


**Figure 4.3 The First Section of the Harbor includes the Yard**

> ***Layout Drawings and Planning***
>
> *It is hard to plan anything if you do not know what you are planning for. In this chapter, I have relied heavily on the drawings I have of the harbor track. While the drawings are very detailed and to scale, it is not necessary that your drawings be as good. Pencil sketches, with the relative locations are normally good enough. The primary purpose is to know where to locate the Nodes, and what to connect to what. Make sure that your drawings are sufficient for that. Of course, you are probably right next to the layout, so you can look at it anytime.*

Moving on to the next section in Figure 4.5, I see less functions, so I am hopeful that I will not need as many boards. I count 10 turnouts, but four of those are crossovers. Maybe I can either mechanically link them to one machine or I can drive both in one crossover with the same driver. That would reduce the number of drivers to 8 which would be fine for an SMD8 daughter board. There are 7 block detectors on this section, so that would fit in a BOD-8. Those two boards would plug into a Tower-LCC. That leaves the 4 track relays. Possibly a Signal-LCC would provide those interfaces. This section is one Tower-LCC with one BOD-8 and one SMD8. The second Node will be either a Tower-LCC or a Signal-LCC with a SMD8, or maybe nothing, if the relays are 5 volt and less than 25 ma.

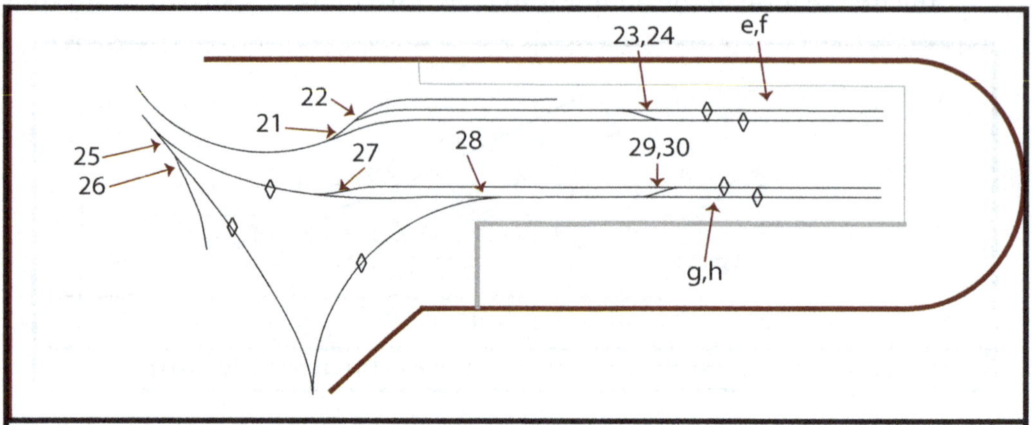


**Figure 4.4 The Peninsula section is relatively Simple**

The final section shown in Figure 4.5is also a simple set of track. Only 8 turnouts, 5 block detectors and 2 track relays. Maybe one Tower-LCC with a SMD8 and a BOD4 …that would take care of everything except one block detector. Oh, RR-CirKits does have a single block detector — that could be used with one of the GPIO pins on the BOD-4CP with some complication. Maybe it is better to just use a second Tower-LCC or a Signal-LCC. The first Tower-LCC would have a SMD8, and a BOD-8, which would take care of everything except the 2 track relays.

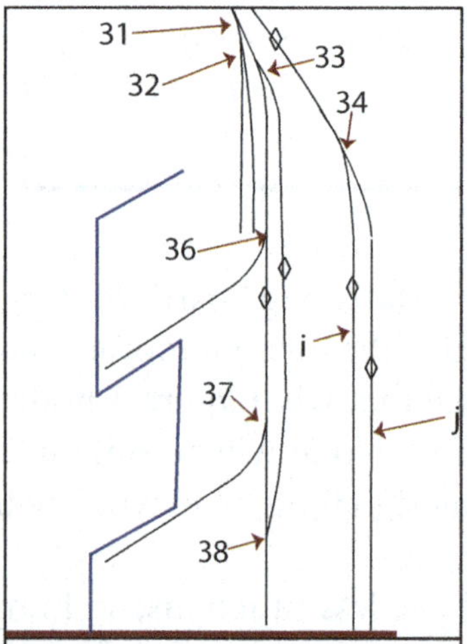


**Figure 4.5 This Section is the First to Implement and the simplest.**

Another solution would be to move the top two turnouts to the second section and use the extra part of the second Node for those turnout machines. Then this section would be just one Tower-LCC with a SMD8 and a BOD8.

I noted before that this section will be the hardest to implement because it is over the stairway. I have designated this as the first section, because it is the simplest.

The next step in allocating the boards is estimating the physical location of the boards. This step will further define which daughter boards are in which Nodes. The general process is to use the section drawings and estimate a position for each Node. Probably the Nodes will be spaced roughly equally in the section, assuming that the functions are evenly spaced. Then draw lines from each function to the nearest Node, keeping track of the number of each function that is connected to a Node. For example, connect turnouts to the nearest Node until you have 8 connected for an SMD8. Then decide if you want another SMD8, or something else in this Node. Then continue to draw lines for each function up to the limit of each Node and each daughter board. Figure 4.6 shows the result for section 1. The drawing is very messy, but pretty is not the point. This is to allocate the functions to boards on the Nodes. In this drawing, the dashed lines are the daughter board connections, with red lines from turnouts to SMD8 and Blue lines from block detectors to BOD-8 boards. The required boards for the harbor are summarized in Table 4.1. The Signal-LCC boards are only partially used, but the anticipated signaling will use more of these two Nodes. Even without signaling, the harbor has 6 Node

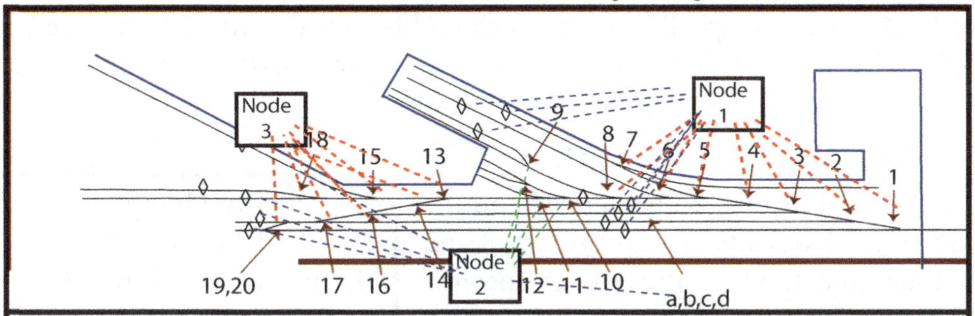


**Figure 4.6 Section 1 "Rat's Nest" of Connections to Functions**

and 10 daughter boards. This also does not include a control panel, which will have many more inputs (primarily push-buttons) and LED indicators. A quick estimate of the inputs and outputs would be one push-button per route (some turnouts will require two) one push-button per track relay, and an LED for each

**Table 4.1 Board Allocation for the Harbor**

| Section | Node | Board | Function | Section | Node | Board | Function |
|---|---|---|---|---|---|---|---|
| Section 1 | Tower-LCC | | | Section 2 | Tower-LCC | | |
| | | SMD-8 | Turnouts 1 - 8 | | | SMD-8 | Turnouts 21 - 30 |
| | | BOD-8 | Blocks 1 - 8 | | | BOD-8 | Blocks 14 - 20 |
| | Tower-LCC | | | | Signal-LCC | | |
| | | SMD-8 | Turnouts 9 - 12 | | | SMD-8 | Track Relays e - h |
| | | | Track Relays a - d | | | | Turnouts 31, 32 |
| | | BOD-8 | Blocks 9 - 13 | Section 3 | Tower-LCC | | |
| | Signal-LCC | | | | | SMD-8 | Turnouts 33 - 38 |
| | | SMD-8 | Turnouts 13 - 20 | | | | Track Relays I, j |
| | | | | | | BOD-8 | Blocks 21 - 25 |

push-button plus one LED for each block detector. Roughly 33 routes, and 10 track relays equaling 43 push-buttons; If the turnouts are controlled by 33 buttons, and only 5 are routes, then the other 28 need two LEDs for direction —

that would be 28 doubled plus 5 routes plus 10 track relay indicators. That is 71 LEDs. Total estimate is 119 I/O pins for a full control panel. If all of the Nods were Signal LCC boards, that would require 5 Nodes. Probably the Signal-LCC would not be satisfactory for all of the I/O, so probably a Tower-LCC would be added. To cover all of this in this book would be long and repetitious, so instead, I intend to subset the harbor functions to demonstrate the actual assembly of a working prototype.

# CAN Bus Limitations and Design Needs

Before we leave the harbor, we should look at some of the limitations of the CAN Bus. Two specific concerns should be noted, although they may not apply to the harbor design. These are the maximum length of the network cable segment and the power required for the Nodes.

## CAN Bus Cable Length

The length of the CAN Bus depends on the number of Nodes attached to a segment and the total time delay. Timing sets an absolute maximum length of 1000 feet, but voltage loss limits the length as well. Each Node requires a small current from the CAN Bus, thereby reducing the total length. Since new technology has reduced the load of the receivers in the Node, the length is dependent on the specific CAN chips. Guidelines for estimating the length state that each Node reduces the total length by 20 feet. Also, the recommended cable length between Nodes is stated as one foot. Given these two restrictions, a CAN Bus can only have 48-49 Nodes. With newer technology CAN Receivers, the load may be less, so the 20 feet per Node may drop, but the overall length is limited by the timing of the bus, so it is probably best to stick with these estimates.

> **Really, One Foot?**
> A very detailed analysis of the CAN bus shows that a mixture of long cables and very short cables can lead to some Nodes being locked out. Since our networks include short segments in control panels and much longer ones for the layout length, sticking to this minimum is correct. On our test network where all the segments are short, we can relax the rule.

Increasing the length with bridges and repeaters is beyond the scope of this book. A repeater only addresses the voltage drop part of the length issue, so the timing is still an issue. Bridges actually create another CAN Bus segment which has the same length restriction as the original segment. Potentially a bridge could stretch a CAN system to 2000 feet, but it does have some time delay in passing the messages between segments.

In our case, the harbor area is roughly 16 feet square, and the bus will probably start at the entry to the harbor, then go to the other side, loop out maybe halfway to the peninsula, then go down to the bottom end. I would estimate that to be 16 feet plus 2x12 feet, plus 16 feet. That is about 56 feet with 7 nodes — we must count the USB buffer as a physical node — at 20 feet each. The total of 196 feet is well within the maximum of 1000 feet.

## Power Distribution

In an earlier chapter the CAN Bus was shown with wires to provide power to the Nodes. When we set up the minimal LCC network, we used a PowerPoint from RR-CirKits to power the CAN Bus and the Nodes attached to it. The PowerPoint provides 500 ma. to each RJ45 connector, which means that it should be in the middle of the network. To get this power, it has a wall power adapter to plug into the wall outlet. This then connects to the PowerPoint to provide power on the CAN Bus cable.

Additional power information is printed on every board — it is required to mark each board with the power used or the power provided by that board. Unfortunately, the manufacturer cannot know the power required by other devices connected to the board. For example, the Signal-LCC has 16 LED terminals. If, in the unlikely case, all of these are On, somewhere between 160 and 320 ma. could be consumed. The board is marked as "20 ma + Load." This is correct for the steady state case, and could be as high as 340 ma. Considering that the PowerPoint only provides 500 ma. This one node will nearly use all of the power available from one side of the PowerPoint.

Many Nodes have terminals for additional power input. Daughter boards like the SMD8 need more voltage than the 5V supplied by the LCC Nodes. Using these power terminals reduces the current demands on the CAN Bus. In the example for the Signal-LCC, the power for the LEDs can be supplied by an additional power supply.

The distance between the power and the Node is provided as a "rule-of-thumb" in the LCC tech notes. This distance, 20 feet, is intended to ensure that the voltage drop due to cable resistance is not excessive. Ordinarily, this would only be a problem if the Node on the end is using the maximum current. In a distributed system, the current used will be spread over several feet. If you want to, you can calculate the voltage drop more precisely.

The most important limit is the maximum current on the cable. Using a PowerPoint as a supply makes this an easy limit, since the PowerPoint limits the output on one of its connectors to 500 ma. Note that the total available from the PowerPoint is 1000 ma. The proper place to put the PowerPoint is in the middle of several Nodes. Then each cable can supply up to 500 ma.

How many power supplies do we need? That depends. All of the estimates above influence the decision. As you build up the network, you may start to Experience strange results. If you remove a Node from the side of the PowerPoint that is failing, does that make it better? If so, you may have hit a limit of current. Try moving the a PowerPoint or adding another. The estimation of power required is not exact, so, your results may vary.

# A Manageable Subset of the Harbor

After playing with the harbor and seeing how complex it is, I mentioned that we would look at a smaller subset to work with in this book. There are a few functions that I want to test before committing to the task of implementing LCC for the harbor. The primary functions that I want to test are the route switching, the track power relays and eventually, the signaling. Along the way I intend to develop some experience and some documentation tools to aid the task and to provide a guide for any future maintenance.

The subset I chose is shown in Figure 4.7. This subset is representative of the harbor yard. It might be thought of as a 5 track classification yard, or a smaller part — the two lower tracks — can be used to demonstrate signaling for a passing siding.

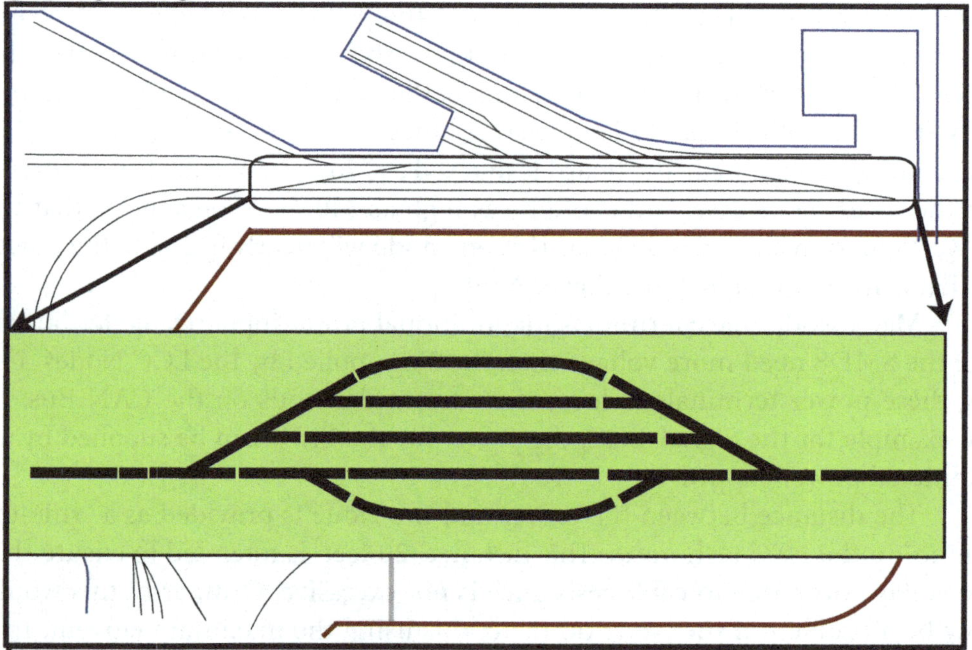

**Figure 4.7 A Simpler Subset to Demonstrate the Needed Functions**

Other possibilities include:

- A grade crossing with the short segment of track on the left end;
- Independent route selection for both ends of the yard;
- Portability because the demo layout can be split in the middle;
- Demonstration of different turnout machines
- Fully open wiring to show how it works.

These additions demonstrate the advantages of LCC to distribute the functions. Splitting the track plan in the middle is no problem, because the CAN Bus is just one cable. Two additional wire pairs are used for track power and Turnout power, making a total of one cable and two two-conductor connectors at the center.

As before, we need to count the Functions to determine how many Nodes and what types to use. Using the same scheme as before, the functions are enu-

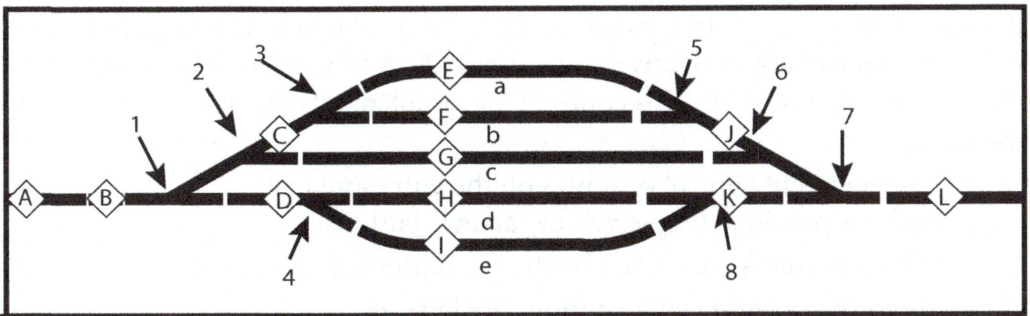


**Figure 4.8 Counting Functions Determines How Many Nodes.**

merated in Figure 4.8. Eight turnouts (numbered), twelve occupancy detectors (capital letters in diamonds) and five track relays (lower-case letters) are the functions needed for this simple track plan. A small complication comes up because I am using Kato track to make the layout. I want to use some of the Kato turnouts with the single-coil solenoids, so I will need a couple of H-bridges for that. The other Kato turnouts are powered with Tortoise machines. A SMD8 will be needed for the stall-motor machines; another one will be needed on the other end (the right end) to power the stall-motor machines there. Since I have two SMD-8 boards, I will use the extra outputs on them to Power the track power relays. Adding occupancy detectors and the necessary power and USB buffer, the result looks like Figure 4.9. The daughter boards support the special I/O but the H-bridges and the additional occupancy detectors are wired directly into the Tower-LCC Nodes. This gets us to about the same point in planning that we achieved in the overall harbor plan. Additional power for the turnout machines

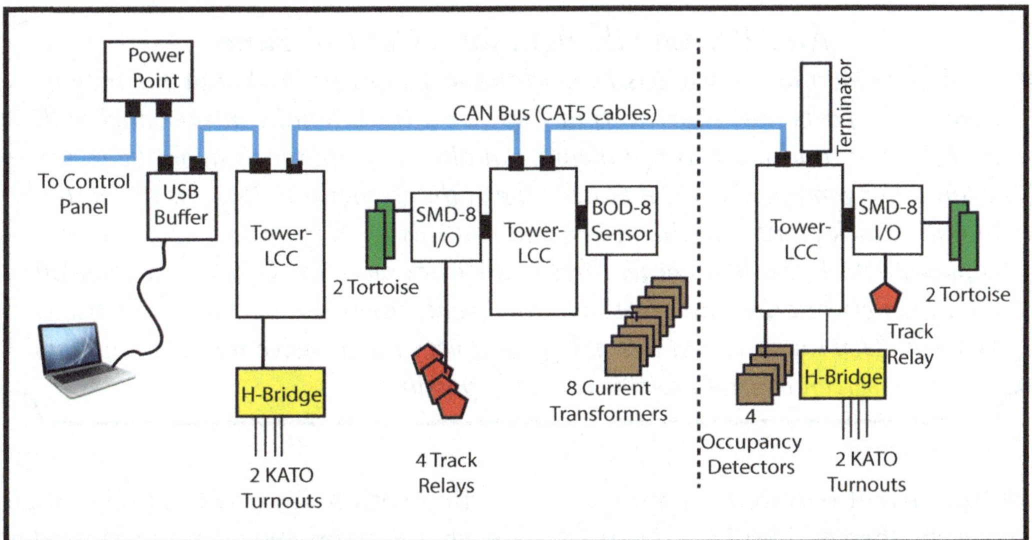


**Figure 4.9 Nodes and Daughter Boards Planned for the Demo Layout**

and track relays is not shown in Figure 4.9. A 12 volt supply will be attached to the SMD-8 boards and the H-driver boards to power these features.

The next step is to create a control panel design for the little layout. The primary devices on control panel are LEDs and switches. This suggests that a control panel will not need any daughter boards. A first draft of the control panel is shown in Figure 4.10. This control panel implements the features previously discussed:

- Independent control of routes on the entry and exit sides is provided by the yellow "arrow" buttons;
- Track power is controlled by the red buttons in the center;
- Both the red and yellow buttons are lit from behind with LEDs — this saves space on the panel;
- Occupancy detection is shown by the blue LEDs.

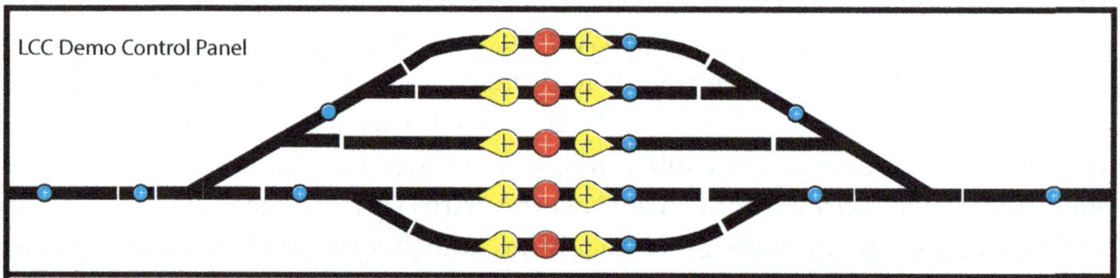


**Figure 4.10 Many LEDs and Push-buttons are needed in a Control Panel**

Enumerating the I/O needs for this panel is somewhat sobering. There are 27 LEDs and 15 push-buttons for a total of 42 I/Os. With the Tower-LCC, that means three Nodes and a lot of wires.

Since the control panel is powered by the cable, maybe a quick estimate

***An Alternate thought for a Control Panel***

*I will not elaborate, since it is beyond the scope of this book, but the magnitude of the parts and effort to [put together a traditional control panel with push-buttons and LEDs is significant. An alternate approach might be to use a computer generated control panel. One of the strengths of JMRI is the ability to make track plans and then to embed controls for turnouts, sensors like occupancy detectors and other features in the track plan. JMRI acts as a Node on the LCC network and then can send the Events necessary to control the track. With modern touch screen technology and very inexpensive micro-computers a touch screen control panel is certainly possible.*

of the current required is appropriate. Starting with 3 Tower-LCC nodes at 20 ma. each, then adding 27 LEDs at 10 ma. each, the estimated current required is

27x10 +3x20 = 330ma. I can ignore the push-buttons since they do not require current except when pressed. I also have made this the very worst case, since all the LEDs will never be on at once and the 10ma current for each LED is probably high. 330 ma. is still below the output of a PowerPoint, so one PowerPoint should be adequate.

## An Alternate Control Panel Design

Using the Sampling feature explained in the last chapter, significant savings can be gad. 15 of the LEDs can be combined with the 15 pushbuttons. This reduces the number of I/Os to 27, which can be provided by two (2) TowerLCCs instead of three. This savings will be the design used in the future chapters.

## Summarizing the Subset

A simpler track plan provides most of the features necessary to implement LCC for the Belmont Shore harbor. Testing the design with this subset will involve a full LCC network implemented with:

1. Five Tower-LCC Nodes;
2. Two SMD-8 daughter cards;
3. One BOD-8 block occupancy daughter card with the current Transformers;
4. One PowerPoint to provide power;
5. One USB Buffer to interface with a laptop for configuration;
6. Push-buttons (lighted) and LEDs for the control panel;
7. Cables and wires to connect everything.
8. Track and control panel, along with all the track power and turnout machines.

While all this is being obtained and assembled, we will start the next chapter with some documentation to ease the configuration process.

# Chapter 5 – Putting It Together

Now that we have the hardware identified and know what we are getting, it is time to consider assembling the demonstration layout. I do not intend to detail the laying of track, the soldering of components, or mounting the cards. I will discuss some of the tips and tricks I used for assembling this demonstration layout. As always, individual preferences may vary. You may not like how I did this, so do it your way … I did it my way. Remember, the goal is to demonstrate LCC features. Since the first use of this layout, some features have changed, but the basic layout will be re-purposed to demonstrate new features.

One of the major features of LCC is the reduction of long wires between ends of the layout. One source of multiple wires is th[1]e control panel. Using LCC, a control panel, no matter how complex, can be connected with just one cable. The LCC network CAT5 cable is all that is needed. This also means that the assembly and initial test of a control panel van be separated from the rest of the layout. In this example, the control panel will be assembled in the last part of this chapter.

## Track and Power

### Documenting the Assembly

Before starting on any assembly, I like to have the documents at hand to guide the assembly. In this case, that is the manuals for each board type and the plans we did in the last chapter. I suggest that you need more than that before you start connecting wires. Each Node has 16 I/Os that will be connected to something, and that will need to be referenced in configuration. My first step is to define these connections so that I know how to wire everything. I created a table for each Node that defines the Node ID, the daughter cards attached and the I/Os by number. The I/Os relate to the Line numbers in configuration, so I am starting to gather the data needed for configuration. I can't fill in all of the

| **Table 5.1 – Track Center Node** | | | | | | |
|---|---|---|---|---|---|---|
| Node ID | | Description | | | | |
| | | Line # | Function | Wire Color | Configuration |
| | | **Track Board Center** | | | |
| | SMD-8 | *1* | | | |
| | | *2* | | | |
| | | *3* | *Tortoise* | | *Out - Active Hi* |
| | | *4* | *Tortoise* | | *Out - Active Hi* |
| | | *5* | *Block Relay* | | *Out - Active Hi* |
| | | *6* | *Block Relay* | | *Out - Active Hi* |
| | | *7* | *Block Relay* | | *Out - Active Hi* |
| | | *8* | *Block Relay* | | *Out - Active Hi* |
| | BOD-8 | *9* | *Blk Detect* | | *In - Active Lo:* |
| | | *10* | *Blk Detect* | | *In - Active Lo:* |
| | | *11* | *Blk Detect* | | *In - Active Lo:* |
| | | *12* | *Blk Detect* | | *In - Active Lo:* |
| | | *13* | *Blk Detect* | | *In - Active Lo:* |
| | | *14* | *Blk Detect* | | *In - Active Lo:* |
| | | *15* | *Blk Detect* | | *In - Active Lo:* |
| | | *16* | *Blk Detect* | | *In - Active Lo* |

squares now, because I need to wait for the final boards. The format is shown in Table 5.1 including the missing data slots. The first addition will be the Node ID vertically in the first column. Just before the Node is screwed down in the center, this can be filled in. Additional identification of the Function column will be helpful during configuration. Each Node has a table like these to guide the assembly and configuration. I used Microsoft Excel® but it is your choice how to make the tables. Even if you draw them by hand, they will be useful. Each Node is planned on these tables. As shown, the design that was started in the last chapter is now contained on these sheets. You can see that the Nodes and daughter boards are on these tables. Looking foreword, Most of the data in these tables should be transferred into the Node memory during configuration. Each Node has a space to identify how it is connected.

## Initial Assembly and Wiring

I will not detail the assembly of the devices. I don't want to insult or bore the readers. If you are this far, you have worked on tracks, turnout machines, and some electronics. The next few paragraphs will only describe the top-level assembly. Where something is unique in my eyes, I will detail that. Most of this was already complete in the previous use of this layout.

Track-work and track power are personal choices In this demonstration, I opted for snap together track and DCC track power. This made the track go faster. Since I wanted to separate the halves for transport, I used expansion track segments in the center. I also used mounting tape to fasten the track down. While it is not as secure as other means, it is adequate for this purpose. After a few times taking it apart, some of the tracks started to break loose at the joint between the halves. I used hot glue to fasten them back down.

Mounting the Nodes and devices uses stand-offs through the mounting holes provided. Daughter cards plug into the Nodes, even though that may not be the best choice for a permanent layout. Future installations will require a different choice, since the latest daughter boards do not plug directly intro a Node. Since I was going to be transporting it to different places, I used cable clamps to hold down the wires. Liberal use  keeps the wires all under control.

The second thing to note is that the Nodes are all wired with ribbon cable. Each Node has two 10-pin IDC sockets. Ribbon cables (the multi-color strips in the control panel) are crimped onto the connectors and then split between the various devices. As I said before, don't waste your time trying to crimp these connectors with anything but a crimper like that in Figure 5.1. Check your favorite on-line store — this crimper was less than $20.

I also mentioned earlier that I prefer color-coded cable, because I don't have to count conductors. Even if I used the cables from RR-CirKits, I would

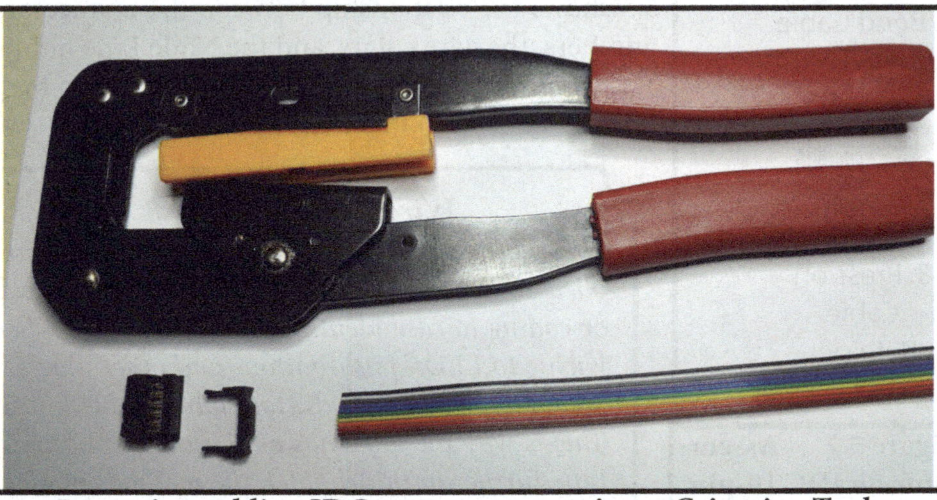

**Figure 5.1 Assembling IDC connectors requires a Crimping Tool**

have to crimp the connectors on. Connector assembly is shown in Figure 5.2. The connectors as shown are actually three pieces, but typically th left two are together. Step 1 is to slide the ribbon cable into the connector, making sure that the wire colors are in the proper order. I will say more about that shortly, but if you use this as a guide, it should come out like the Tower-LCC.

To complete step 1, insert the connector into the crimper (the crimper is slotted to accept the key), gently squeeze the handles to contact the connector, then squeeze until the gap between the parts closes.

Step 2 is to bend the cable back over the top of the connector. Squeeze it down to make sure the cable is tight against the top, then carefully fit the cable retainer over the top and down the slots on the sides. If you have it all aligned correctly, the retainer should snap into the side and stay, Trim any excess off the side, and you have the ribbon cable for the node I/O.

I use a hobby knife or a razor blade to separate the conductors, then I use a good quality wire stripper on the conductors. The big question at this point is which wire goes to which internal Line? I have nicely color-coded wires, and I have connector sockets on the Node, so I need to associate the wires with the internal Line numbers.

The first thing to notice is that the Node (a Tower-LCC) has two sockets, labeled "I/O-1" and "I/O-2." The lines connected to I/O-1 are 1 to 8, and I/O-2 has 9 to 16. Now to connect the wires with the Lines. Figure 5.3 shows the relationship between the connector pin numbers, the wire colors, and the Node Line numbers. This step — filling in the table — can either be performed as the connections are made, or it can be

Steps for IDC Connector Assembly

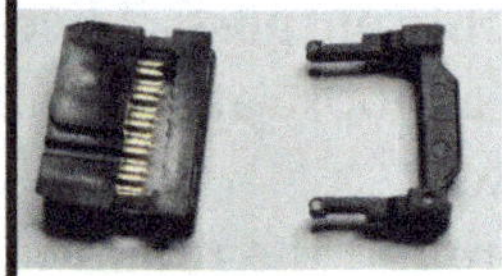

Parts

1. Insert Cable and Crimp

2. Bend Cable Over Top

3. Press on Cable Retainer

**Figure 5.2 Assembling an IDC 10-pin connector to A Ribbon Cable.**

### *Wire Color — Note 1*

*The cable for the RR-CirKits connectors is supplied in a plain gray wire. I wanted to use color coding to ease identification of the lines when wiring to LEDs, push-buttons and other places. I bought roll of 10-conductor ribbon cable to use. This is why the table has a wire color column. A little further in this chapter, the methods of making these cables will be described.*

planned ahead of time. Either way, the table will document the wiring that connects the devices to Lines in the Node.

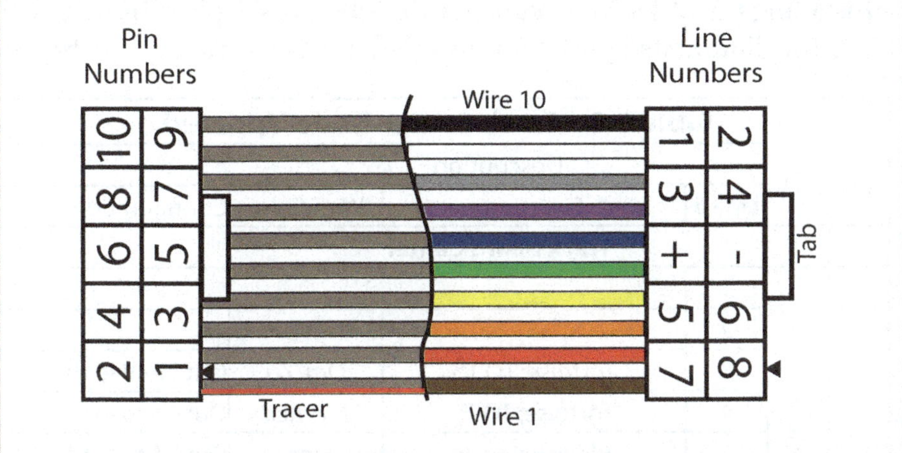


**Figure 5.3 Connector Pin Numbers do not match Line numbers**

Wiring the SMD-8 daughter board twists the wires around again. On the SMD-8 connector, the last two wires — 9 and 10 — are connected to Motor 1, the next two are connected to Motor 2, and so on. I connected each pair to a PC connector for the Stall Motor and then I can simply turn over the connector if it is backward. The wires for the track power relays are not so easy to change. Using the SMD-8 to power a relay must be more specific during the wiring. Updating the original Table for the Track Center Node results in Table 5.2 on page 84.

During the wiring of the Nodes, it was discovered that the TowerLCC on the Lef6t end of the layout only required four lines. Looking forward to installing signaling, this Node was changed to a SignalLCC for that use in a future chapter.

Each Node has a table like these above. As the Nodes are wired, the tables are a guide, and are updated as needed to match the final wiring. At the completion of the wiring, these tables will help define the configuration. Including the wire colors will make tracing the signals easier if the layout has to be maintained.

# Control Panel Assembly

There are many ways to make a basic control panel. Look on the Internet and choose one. I have a simple method, although it depends on computer drafting (drawing) tools for the face of the control panel. A very effective panel results from the drawing. The result is a drawing like Figure 4.10 on page 76. Starting from that drawing, the control panel is a frame like a shadow box frame or, in this case, just a simple open box with a plastic sheet on top. I needed a 1/16 sheet because of the buttons I choose. On this sheet I glue a laminated print of the framing, then drill holes as needed. The Nodes – two TowerLCC+Q boards – are mounted on the sides of the box.

Installing the LEDs for occupancy detection is simply a 5mm hole to glue in a LED. The illuminated push buttons (they are called tactile switches because

**Table 5.2 – Track Center Node Updated**

| Node ID | | Description | | | |
|---|---|---|---|---|---|
| | | Line # | Function | Wire Color | Configuration |
| | | | **Track Board Center** | | |
| | SMD-8 | 1 | | | |
| | SMD-8 | 2 | | | |
| | SMD-8 | 3 | Tortoise; TO Tb | *Org/Yel* | Out - Active Hi |
| | SMD-8 | 4 | Tortoise; TO Tc | *Brn/Red* | Out - Active Hi |
| | SMD-8 | 5 | Block Relay- B | *Black* | Out - Active Hi |
| | SMD-8 | 6 | Block Relay- B8 | *Gray* | Out - Active Hi |
| | SMD-8 | 7 | Block Relay- B7 | *Yellow* | Out - Active Hi |
| | SMD-8 | 8 | Block Relay- B6 | *Red* | Out - Active Hi |
| | BOD-8 | 9 | Blk Detect- B6 | | In - Active Lo: |
| | BOD-8 | 10 | Blk Detect- B7 | | In - Active Lo: |
| | BOD-8 | 11 | Blk Detect- B8 | | In - Active Lo: |
| | BOD-8 | 12 | Blk Detect- B9 | | In - Active Lo: |
| | BOD-8 | 13 | Blk Detect- B5 | | In - Active Lo: |
| | BOD-8 | 14 | Blk Detect- B2 | | In - Active Lo: |
| | BOD-8 | 15 | Blk Detect- B4 | | In - Active Lo: |
| | BOD-8 | 16 | Blk Detect- B3 | | In - Active Lo: |

they are used primarily on keyboards) are designed to be inserted into a printed circuit board (PCB). So, I designed a PCB to put them in. While I was designing

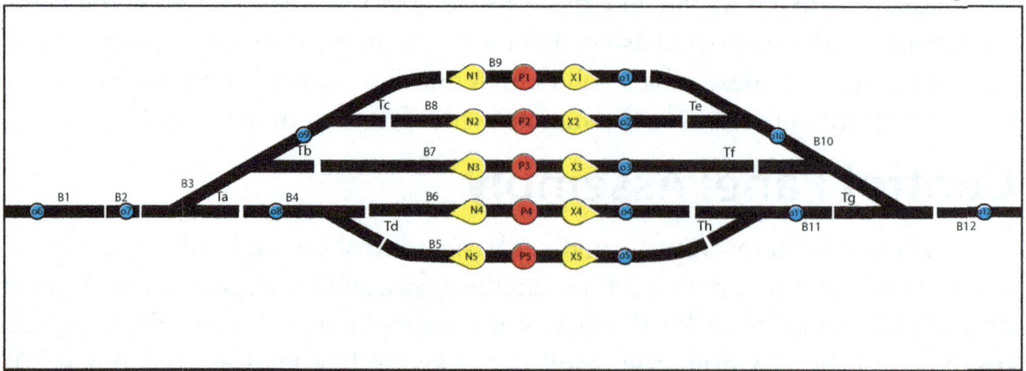


**Figure 5.4 Initial Annotations on the drawing Name the Producers and Consumers to Link to the Nodes**

the PCB, I included 10pin connectors so that I could just use a 10 conductor cable to connect the LED/Buttons to the TowerLCC. This one PCB handles all the I/O from the two TowerLCC boards, providing screw terminals for the LEDs that are not on the central button board. The resistors necessary for the sampling are on the board. The Resistors for the LEDs not on the board are on the board. The PCB is shown in Figure 5.5.

**Figure 5.5 Assembling the Buttons, LEDs and resistors is Easier with a custom PCB**

The annotated Control Panel drawing in Figure 5.4 on page 84 will provide most of the information needed for configuration. Each LED/Button combination will be identified. From the turnout identification, the Line numbers can be referenced to the tables similar to Table 5.2 on page 84. Once the wiring is accomplished, the Track numbers for each sensor will be filled into these tables, providing the information to couple them together. This information can then be entered into the Node configuration. Then the documentation in the Node will be available in the future.

This control panel demonstrates the convenience of LCC. Since the power for the LEDs is in the LCC cable, only one CAT5 cable is needed to connect the control panel to the layout. Additional features like a fast clock display (see Chapter 10) or remote display of signals could easily be added without adding any cables.

## Identifying Producers and Consumers

While the identity of Producers and Consumers seems obvious, it is still an important step to list the Producers and Consumers on the control panel drawing. When each is given a name, the relation between the Producers and Consumers can be better defined. The goal here is to work up to defining the configuration actions before jumping into the deep end. The drawings shown in the next few paragraphs are to illustrate the concepts — I do not think they are necessary for Configuration, since the data is available to jump to the final annotated drawing. In many cases, even that drawing does not need full annotation, since the act of annotating can be superseded by entering the data into the Node directly.

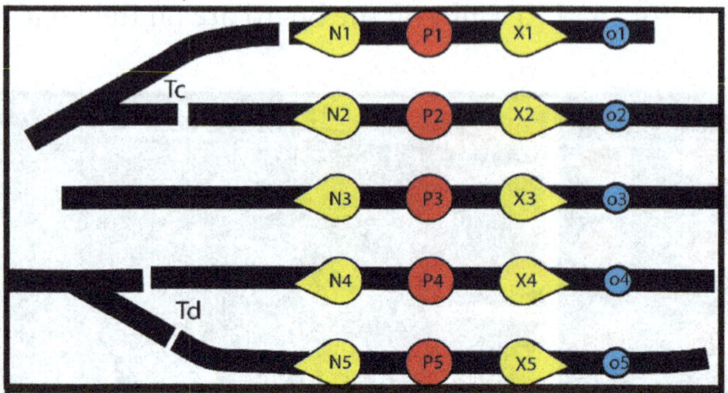


**Figure 5.6 A fragment showing names for reference**

Since everything is either Produced by or indicated on the Control Panel, the fragment of the Control Panel in Figure 5.6 will be used as a reference. The Producers and Consumers on the Control Panel are easy to identify from the functions. In Figure 5.4 on page 84, the Block Occupancy Producers are blue circles numbered o1 to o5. The rest of the functions on Control Panel are Consumers — buttons for the turnouts and track relays. Track relays are numbered from P1 to P5 in red circles, turnouts are numbers in the V of the turnout Numbered Tc and Td. The other turnouts are not shown. Illuminated pushbuttons are in "pointer circles, numbered N1 to N5 and X1 to X5.

Identifying the Consumers and Producers begins with the easy one —the Block Occupancy. In this case, the Producer is the Block Occupancy detector (the BOD8) and the Consumer is the LED on the Control Panel. Almost as easy, the Track Relay buttons are the Producer, but there are two Consumers, the actual relay and the LED on the illuminated button. Since there is only one button, this will be an Alternating button.

Looking at the turnout control, the illuminated buttons labeled N1 to N5 and X1 to X5 are both Producer and Consumer. To indicate on the Control panel which track is selected, pushing one of the buttonsN1 to N5 must turn off all the buttons except the one pushed. This implementation will be left to the next chapter.

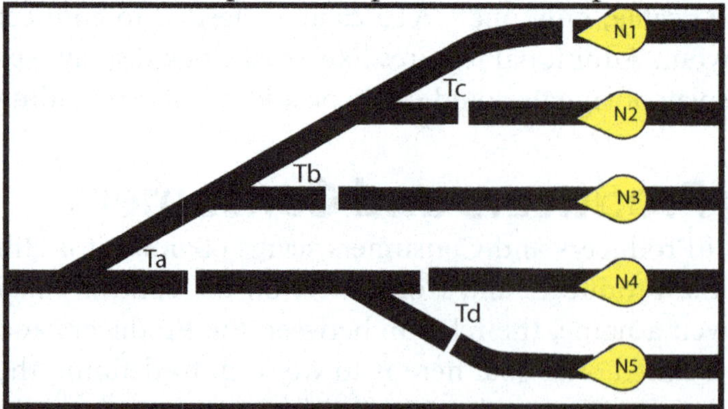


**Figure 5.7 This fragment shows the Left Yard Ladder withe Control Buttons**

Turnouts are naturally Consumer, right? They move the points when commanded. Looking at the fragment representing the left hand in Figure 5.7, some things are clearly controlled by the buttons N1 to N5. Turnout

Ta is not controlled by any illuminated button. Instead, this is where the Cascade feature comes to our rescue. To illustrate, consider pushing button N1. N1 commands turnout Tc to the Normal (through) direction. To get the train to Tc, the other turnouts in the Ladder have to be set correctly. So Tc issues a Cascade event which commands Tb to the Normal position. Then, to complete the path, Tb issues a Cascade event to Ta commanding it to the Diverge direction. This completes the path from the left lead track to the yard track indicated by button N1.When pressed button N2 commands Tc to the Divergent direction, but since Cascade does not specify a direction that Tc is commanded, the Cascades continue to set up a correct path for that track. When all the Cascades are complete, turnout Ta is commanded only by the two Cascade events from turnouts Tb and if. So, while the actual turnout like Tc is only a Consumer, the controller in the Node issues the Cascade to ensure that the path is set up to get to that turnout.

## Connecting Producers and Consumers

The next step is to identify the Producer/Consumer combinations. To prepare for Configuration, the line numbers for each will be added to the control panel drawing. As noted before, the data is contained in the tables like Table 5.2 on page 84. The task at hand is to transfer the Line information from the tables to the drawing as shown in Figure 5.8. The annotation is explained in the

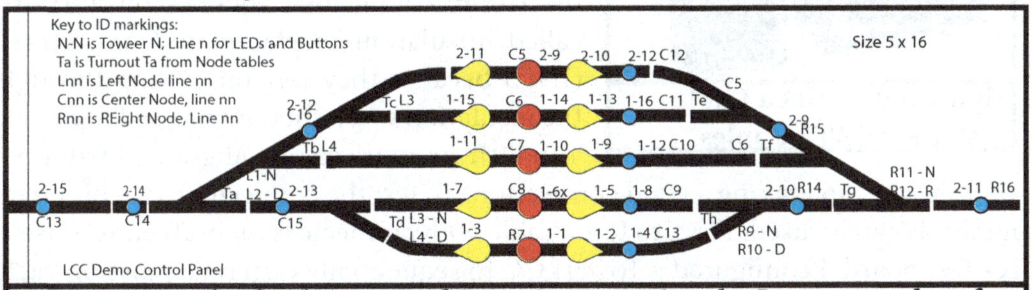


**Figure 5.8 The final A notated Drawing contains the Important data for Configuration**

figure. Use whatever is meaningful to you. The important thing is that it is understood when configuring the nodes.

So, that wasn't too hard was it. The Cascade feature simplified the yard switching so that a route can be selected rather than individual turnouts. This Route Switching is also known as NX switching, which stands for eNtry/eXit.

Route switching has been designed many times for model railroads. Usually the hardware involved is either some kind of relay logic network or a diode logic network. The wiring becomes very complicated and the documentation is difficult. There is a discarded relay network at the Belmont Shore club that demonstrated the difficulty of maintaining these Route Switching networks. When the member left the club, the logic was not well documented. Then when a relay failed, it was impossible to determine what happened. So, Route switching was removed and individual switches put in for each turnout.

# Initial Testing

At least some testing should be done before attempting to configure the nodes. The frustration of assembly issues during Configuration is no fun. Significant parts of the assembly can be tested prior to final Configuration. The first tests verify the network is operational. Then individual sensors and outputs can be tested. This section provides some methods for testing and troubleshooting.

## Cables and Connectors

To validate the network, the cables need to be nearly perfect. If the cables have breaks in some or all of the pins, parts of the network may be inviable while other parts are working well. An important tool if you are working on a network with multiple Nodes is a cable tester for CAT5 like the one shown in Figure 5.9. When the two parts are plugged into the ends of a cable and the unit is turned on, the LEDs on each part will sequentially light. If any LED is dark, the cable has to be reworked or replaced.

Figure 5.9 Get a Cable Tester for CAT5 Cables

The second set of cables to test are the interconnecting 10 conductor cables with the 10-pin connectors. These connectors are called "Insulation Displacement Connectors" (IDC) because they rely on cutting through the insulation when they are pressed together. Too little pressure or mis-alignment will prevent these from making a good connection. To test the connectors, two different methods can be used. First, to test a cable with connectors on both ends, I use a IO Test board. I configured a TowerLCC to sequentially turn on a pulse on each line. Now I can watch each line light up and verify that all of them are connected. The second method would be to attach a voltmeter to each line and check that each line turns on and off. An adapter could be used to match a set of screw terminals to the stripped wires.

## Control Panel Buttons

The next check that can be done is yo see all of the buttons on the Control Panel. Since the TowerLCC has a default to Input On when purchased, no changes to the configuration are necessary. With the Control Panel disconnected from the track section, simply start the Traffic Monitor from the JMRI PanelPro window and push a button. The response should look like Figure 5.10. First verify that the Node ID is what you expected. The NodeID Is the number on the Left. Check your information yo make sure this corresponds with your annotations. Then look at the EventID and check that the button pressed is the one identified in your document. To check this, you will have to look at the Input Events for each line on the board until you see a match with the event from this button.

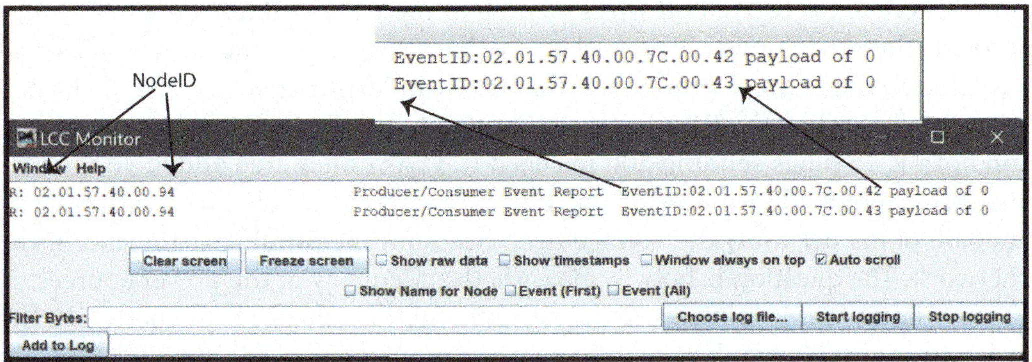


**Figure 5.10 Test Input Producers by Observing Events in the Traffic Monitor**

If you did not get an event (you should have two, one for the press and one for the release), something is wrong in the connection or the button. If the line s don't match, The wires are connected differently then recorded. If the numbers are consistently off by 8, it is possible that the 10pin cable is in the other connector on the board.

## Track Tests

When all the buttons on the Control Panel are checked, connect the track section into the network. The next tests require power for the track and turnouts. Connect these and connect as control station up for the track. Occupancy is the next Input to test. Place a locomotive on a track block and check that events are produced.

Consumer tests use the Trigger feature to simulate the events. Start with the Block relays. Place a sensor on the block to indicate when the block is powered. The sensor can be a locomotive with the headliw1ght on, a lighted passenger car, a voltmeter or anything that will show you when the power is on that block. Now open up the Configuration window that contains the relay, and find the Outputs that activate the Block Relays. Set the Output to Steady and set an event for On and another fore Off. Use Trigger on the Line connected to the Block Relay. It should turn on the track power as indicated by the sensor.

Turnouts are tested like Block Relays, except the direction for On and Off needs to be noted. In configuration this will be important. Also, the implementation for the Kato turnouts is a 100 ms pulse, not a Steady voltage. This needs to be entered before a test on these turnouts.

## Power Adequacy Tests

Power was estimated in Chapter 4. In unusual cases, especially with large layouts, power may be required at multiple points on the layout. Since\w there is no reliable way to predict what power is necessary, it may be necessary to test if sufficient power is injected into the network. Just to review, power in the network is carried on wires 7 and 8 of the CAT5 cable. Some nodes may provide their own power by way of an auxiliary connector, and some nodes may use that to inject more power into the network. Most networks have dedicated power sources like the RR-CirKits PowerPoint. The PowerPoint inje1cts power on both RJ45 plugs sending 500 ma each way. As nodes are4 added yo the network, current load increases. While the resistance of the cable wires is low (0.05 to 0.08 ohms per foot) the voltage drop becomes measurable at the end of the network. The question is how to measure the adequacy of the power sources.

The answer is to measure the current and the voltage on the network. The wires in the cable can be accessed by cutting into the cable, or using a RJ45 Breakout. A simpler, but slightly;y more expensive method is to purchase one or more LCC Voltmeters from RR-Cirkits. This part, shown in Figure 5.11. This device sits in the network, connected with RJ45 connectors. It measures both Volts and Amperes at the point it is connected. As shown, it is indicating 14.17 volts and 96 ma.

**Figure 5.11 Measure Volts and Amperes**

Now the question is where to put this device. First, note that the device has "supply" and "Load" on the back. Put the cable from the Power source in the "Supply" side, and the other cable in the "Load" side. Since the current is distributed between the nodes, the only point where the full current is on the cable is right next top the power source. In Figure 5.12, the current for Nodes 2 to N is provided by the power

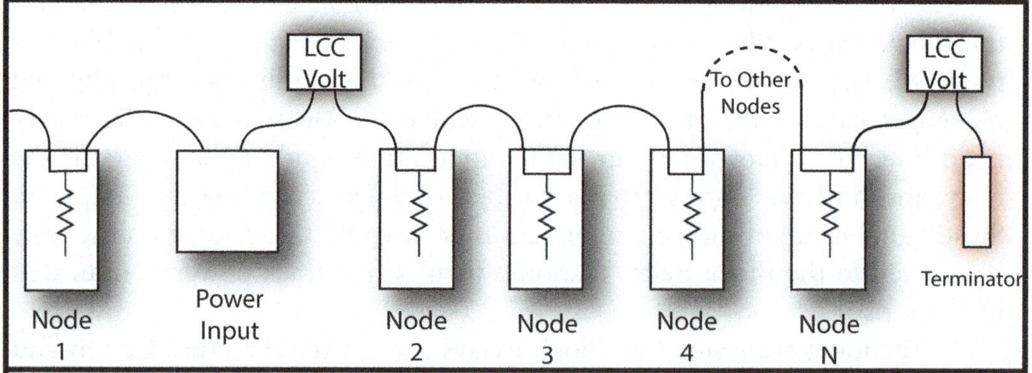


**Figure 5.12 Measure Current at the Sources; Measure Voltage at the End.**

source and will be displayed on the first LCC Voltmeter. Since each cable connection out to the second LCC Voltmeter reduces the voltage, the final voltage will be on that LCC voltmeter. The first meter only shows the initial voltage of the power source and the second meter shows the current into the Terminator.

So how much voltage drop is too much. In my opinion, the voltage for good network performance is 10 to 12 volts. If your network falls bel9ow that, you probably should add some power.

After all these checks are successful, Configuration should go well. Note that no eventIDs were changed in thees tests. There should not be any conflicts between events as a result.

# Summary of Putting it Together

We have designed a simple demonstration layout with some interesting features. Control of turnouts, track power relays and sensors for block occupancy are the main features of this layout. We also have gathered data for route switching which will allow selection of a ladder track with one button push. We also designed the yard tracks to have independent routes on each end, allowing moves in and out of both ends simultaneously.

Documentation has been a theme of this chapter, because it is important to have a record of what you have done. The documentation that I have shown is useful in the assembly of the network and will be critical in the final configuration of the network, which I will cover in the next chapter. Only the final annotation of the Control Panel is needed for Configuration. The tables have been used in the assembly of the track section. In this case, the Control Panel is primarily documented by the design of the PCB. I don't expect anyone to create all of the Tables and fill them in like I show. These tables are primarily to define the data required. If you understand what data is required, it can be entered directly onto the drawing. Then, in the next step, Configuration, the data can be entered into the Nodes for permanent documentation.

The demonstration Layout track sections are shown in Figure 5.13 on page 92. The top two pictures are the left end, and the bottom picture is the right end. One of the first things to note is that I used the connector on the ends of the daughter boards (SMD8 and BOD9.) This would not be acceptable under a layout. Gravity and vibration would probably cause the boards to fall out. Newer revisions of these boards do not have that connector.

As noted earlier, only three small cables connect the right and left ends. These are the LCC Network, track power and +12 volts for turnouts and relays. To show the versatility of LCC, one block relay and its occupancy detector are on the right end, but all the rest are on the left end.

Most of the wires shown in these photos would normally be connected to a control panel or fascia switches. All of that can be consolidated in the single LCC cable that connects to the Control Panel.

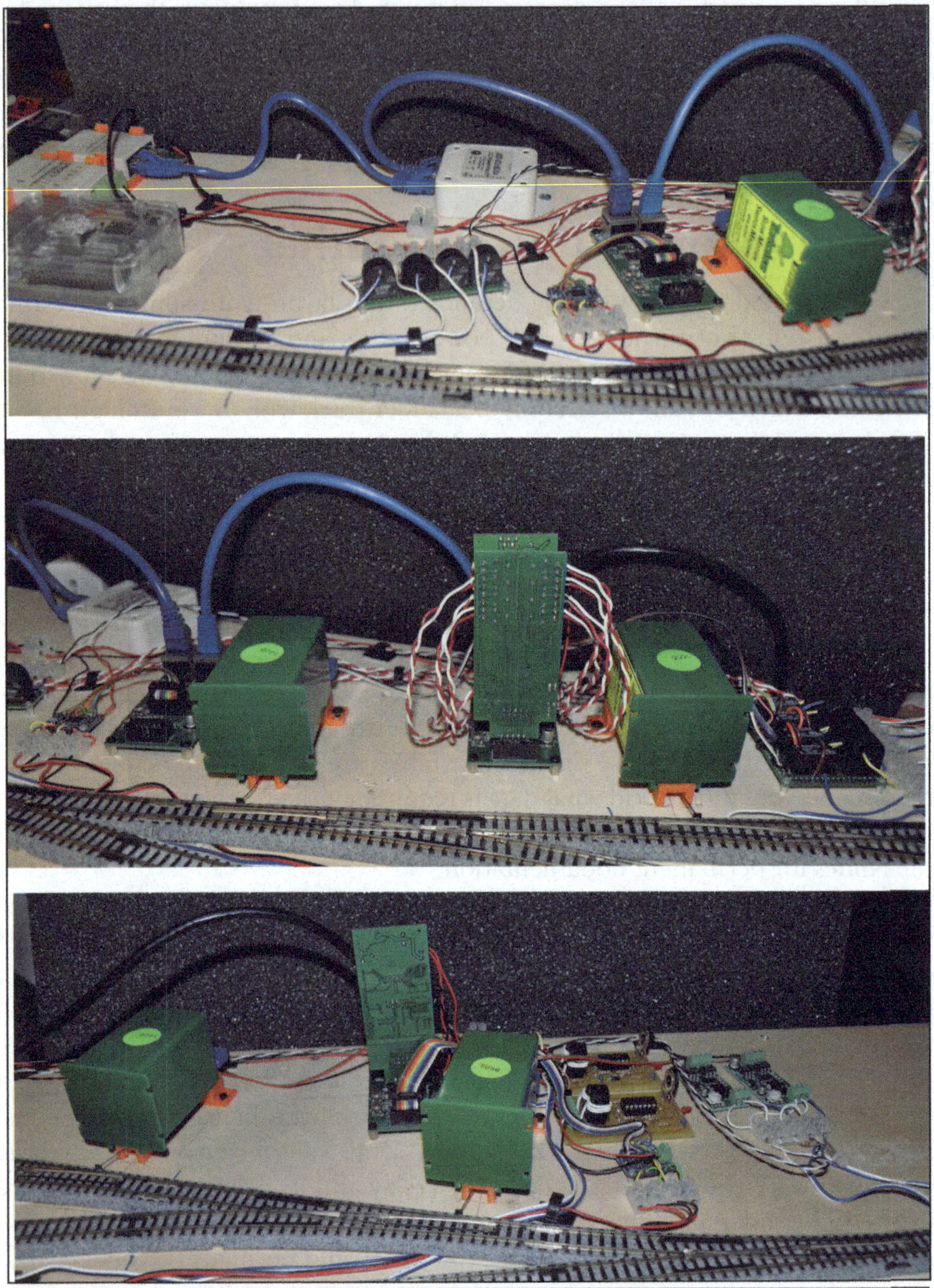

**Figure 5.13 The Track Board Three Nodes for this Layout**

Endnotes
1

# Chapter 6 – Configuring the Network

In the earlier book, I configured the Demo Layout from blank nodes. This time, I am re-purposing the layout to demonstrate some new, powerful features in the TowerLCC. These features are available by way of a firmware update. As noted in the last chapter, I will also use the Sampling feature to reduce the number of TowerLCCs in the control panel. Since this is a new board using a new feature, I will do an initial configuration of the control panel as I test the new board. To do this, I have connected the control panel nodes and button board to the minimal LCC Stack shown in Figure 3.3 on page 43.

After initial configuration of the control panel, the next easiest action is to configure the Occupancy indicators on the control panel, then the power relays. In each of these the EventIDs from the previous nodes will be used to configure the new control panel nodes. The transfiguration for the turnouts will require modifying the nodes already on the layout. The cascade feature is in the latest firmware update for the TowerLCC, so the nodes need to be updated before this. I will update them before any new configuration is attempted.

Configuration will use the tools from JMRI described in Chapter 3. To review that, we first start up LccPro in JMRI with the LCC USB-buffer connected to the computer's USB port. Opening the LccPro program displays the nodes that are connected (see Figure 6.1). Since this is the first time the control panel nodes have been connected, the **Name** field is currently blank. If I had remembered to write down the serial number for these two nodes and remembered which one connect to which connector on the button board, I could add the name. Based on Figure 5.8 on page 87, nodes are simply 1 and 2. To identify

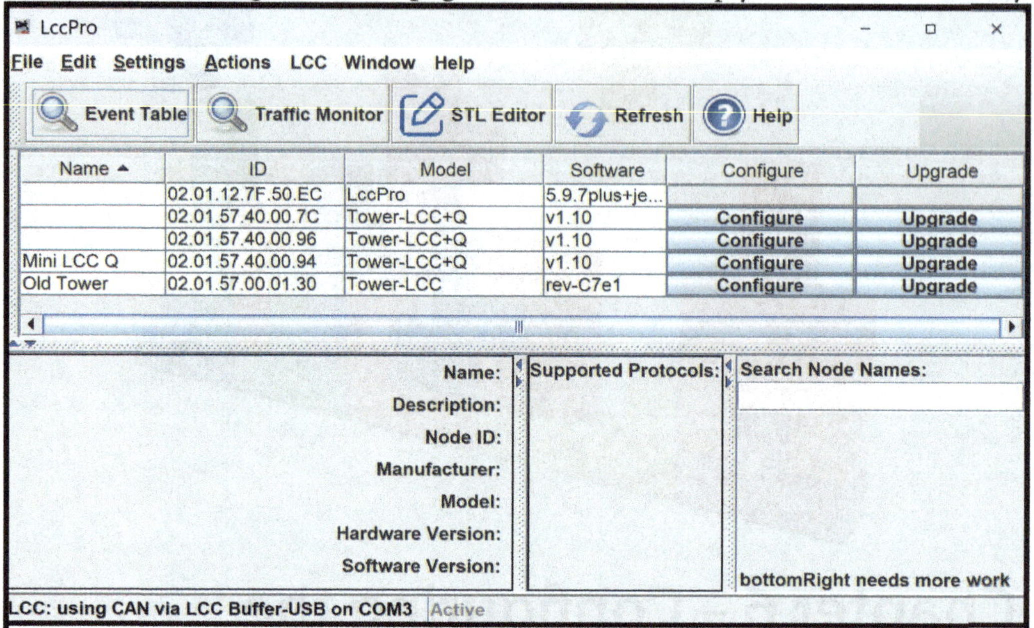


**Figure 6.1 Control Panel Nodes are not yet Named when LccPro Opens**

which is which, I open up the Traffic Monitor and push a button. Since the default initial state of a TowerLCC is **Normal** Input and sends events when On and again when Off, I should see two events on the Traffic Monitor. Associating the Node ID with the button that was pressed will determine the name for one of the two nodes. To test the buttons, I continue to push each button and note the **EventID** for each button. If a button does not cause an Event, I flag it and fix whatever is wrong.

Next, I will test the LEDs in the buttons. I opened up the configuration for Node 1, I set the **Output Function** to **Steady**, then went down to the **Command/Consumers** section and set the Event 1 to the **Action>On**. Then, as shown in Figure 6.2, I selected **More>Trigger** to activate the LED. If it turned on, I moved on to the next one. This method also tests for the simple LEDs used in Block Occupancy.

When I finished testing all the buttons, I started to do an initial configuration on the buttons. The power buttons down the center are simple alternating functions, so that just changes the **Input Function** to **Alternating.** Initially, these buttons should be Active Low so that the LED is lit when the relay is not energized. Initial configuration is limited to setting the Alternating input and lighting the LED. Final configuration will be later in this chapter.

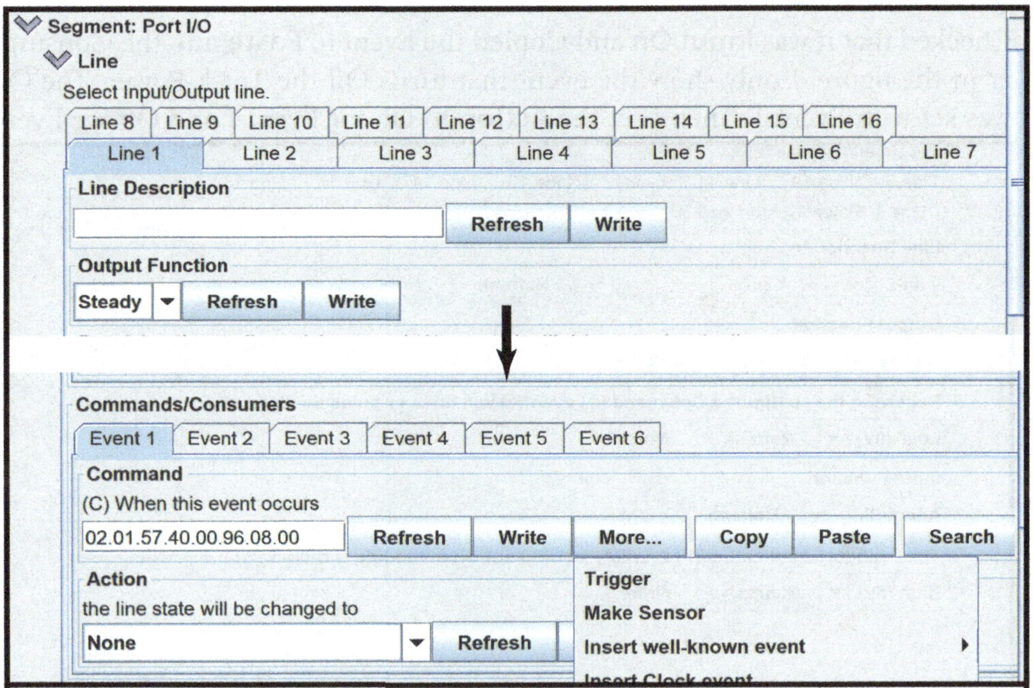


**Figure 6.2 Testing LEDs on the Control Panel**

Using the references on Figure 5.8 on page 87, I selected Line 1and added the name of this Line. Following the red arrows in Figure 6.3, I set the Output type to **Steady**, Set the output to **High** and the Input to **Alternating**. Then I moved down to set up the Input in the **Producer** section. Select Event 1,

### *Lines are High or Low?*

*When setting the High/Low property of a Line, you need to consider what the Line is connected to. First, the Line can be Input, Output or both. Inputs are normally Low, because the μC provides a pull-up resistor when checking for an input. In that case, the High is when there is no signal to pull it Low.*

*Connecting to an Output, the μC connects +5 volts for a High, and Ground for a Low. In this case, the current output must be limited. When connecting to a pushbutton/LED for sampling, resistors must be included in both.*

*Connecting to a device like the SMD8 stall Motor Driver is a connection that does not inherently determine High or Low. The SMD8 changes the direction that a turnout is driven by High or Low. The SMD8 also has a high input resistance, so no resistor is needed. If you want to change the direction that the turnout moves, one way would be to change from Low to High or High to Low.*

*If a pushbutton is desired for sampling with the SMD8 (or similar device) a resistor should be used to limit the current when the pushbutton is pushed. Remember that the Sampling is a very short time. Do not try this with a continuous signal like a toggle switch, because the signal will override the signal from the Node and throw the turnout.*

Checked that it was **Input On** and Copied the Event to **Paste** into the Consumer in the figure, I only show the event that turns Off the Track Power. The On was set with Event 1. Then I set the **Action** to On for Event 1 and Off for Event

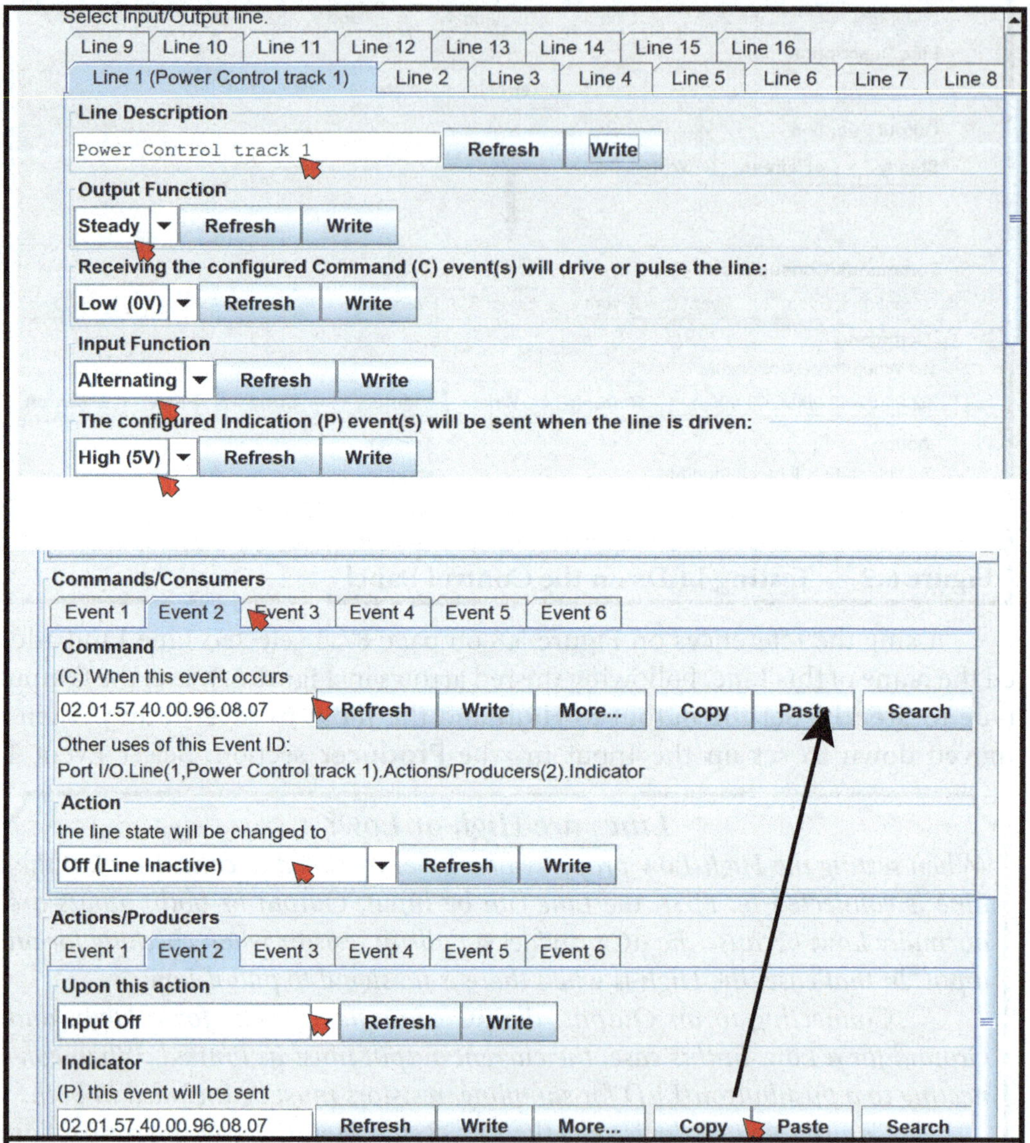


**Figure 6.3 Control Panel Configuration for Power Buttons**

2 Don't forget to **Save Changes**. Repeat these actions for the rest of the Track Power pushbuttons.

The next pushbuttons to setup are the ladder tracks on each side. These are part of the NX switching and should only have one track lit at a time. I use the fact that a pushbutton has two events for each time it is pressed—On when it is pushed and Off when it is released. I use the On event from all pushbuttons in a ladder to turn Off all those LEDs, then turn On the LED connected to the button that was pressed. This setup is shown in Figure 6.4. First, I went to the Producer section for Event 1 and Copied the event for Input On. I then used that Event in each line on that ladder, pasting it into Event 1 for the Consumer, set to turn Off, and Pasting it in Event 1 of the Producer, when it turns On. Then, as shown in Figure 6.4, I went back to each pushbutton and copied Event 2, Input

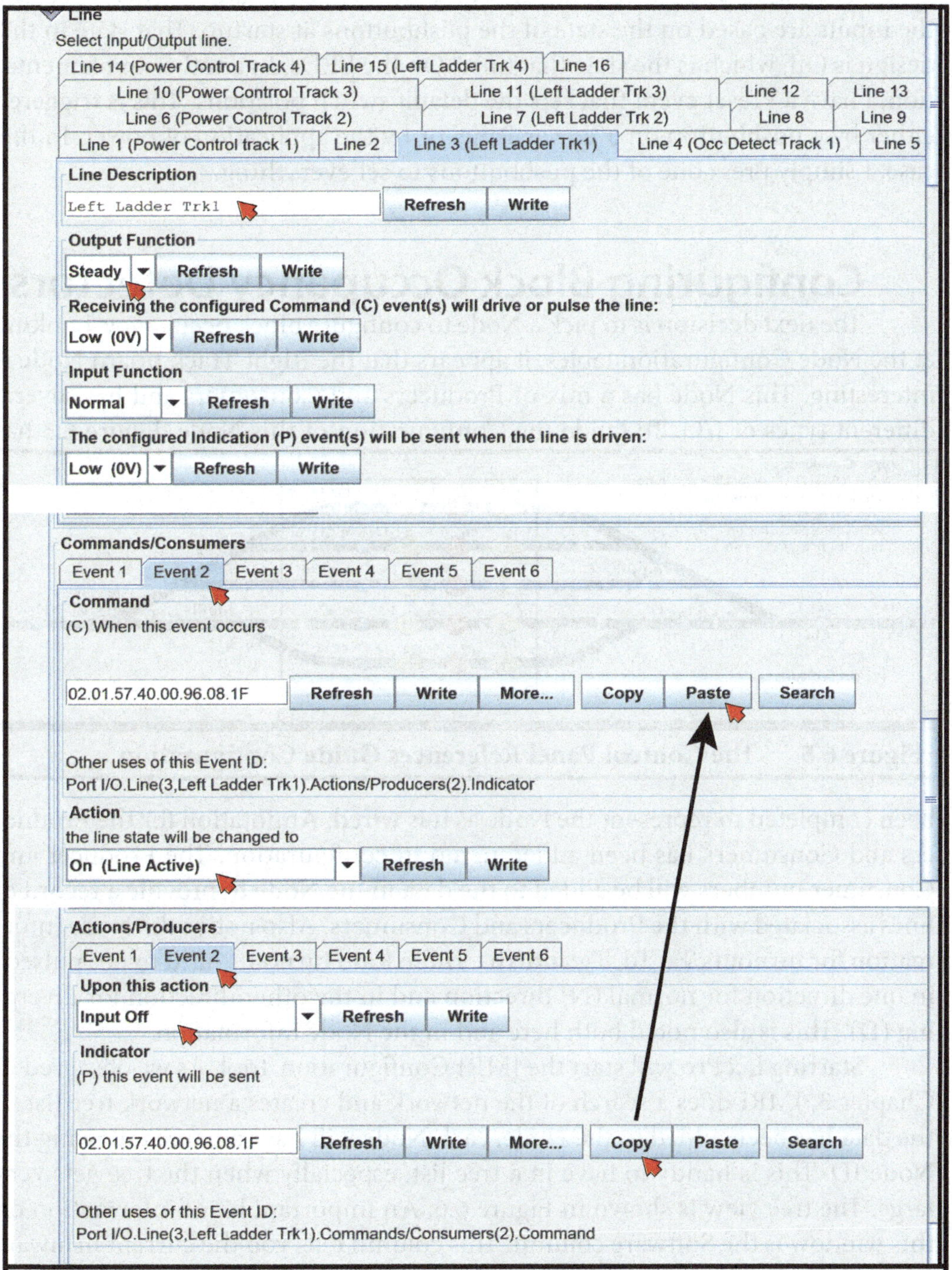


**Figure 6.4 Setup for Ladder Track Pushbuttons**

Off, and pasted that in the Consumer Event 2 for Output On. Only the second step is shown. Repeat for all pushbuttons on the first ladder then do the second ladder with a pushbutton event from that ladder to turn off the LEDs before turning On the one pressed. I tested the configuration by pressing each button on the ladders. Only one was lit at a time, the configuration is correct.

Starting up the panel lights up all the LEDs in each ladder. This is because the inputs are based on the state of the pushbuttons at startup. That state in this design is Off, which is the state that turns On the LEDs. In another implementation, I have a Preset event that sets the default switch positions. This is triggered either by a pushbutton on a Power Panel or by the application of power. In this case, I simply press one of the pushbuttons to set everything.

# Configuring Block Occupancy Detectors

The next decision is to pick a Node to configure block occupancy. Looking at the Node Configuration tables, it appears that the Right Track Board Node is interesting. This Node has a mix of Producers and Consumers and has several different types of I/O. To guide the Configuration of this Node, Figure 6.5 has

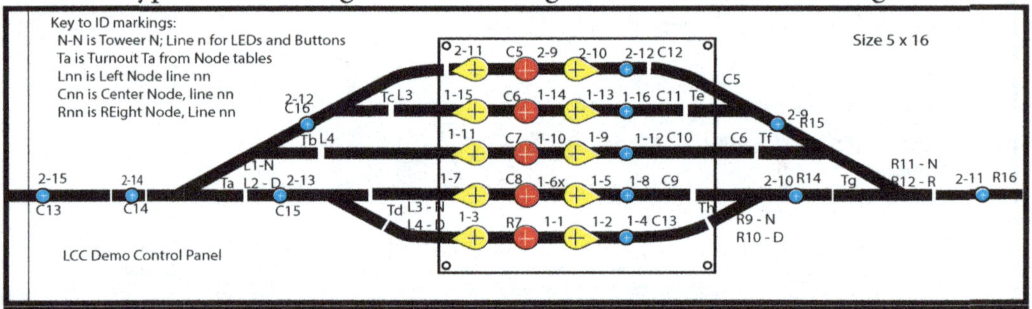


**Figure 6.5 The Conteol Panel References Guide Configuration**

been completed to represent the Node as it is wired. Annotation for the Producers and Consumers has been added to aid in configuration. The Producer and Consumer numbers will be added to the data in the Node to provide a record of lines associated with the Producers and Consumers. Also notice the Node information for turnouts Ta, Td, Tg and Th.. These Kato turnouts need to be "pulsed" in one direction for normal (N) direction and in the other direction for diverging (D). This is also noted both here and in the Node information.

Starting LccPro will start the JMRI Configuration Tool. As we observed in Chapter 3, JMRI does a search of the network and creates a network tree list. If the data has been added to the Node, each Node will have a name following the Node ID. This is handy to have in a tree list, especially when the tree gets very large. The tree view is shown in Figure 6.6. An important bit of information on this window is the Software column. This column tells you the current firmware release installed in the node. Since this is a rework of an existing layout, the dis-

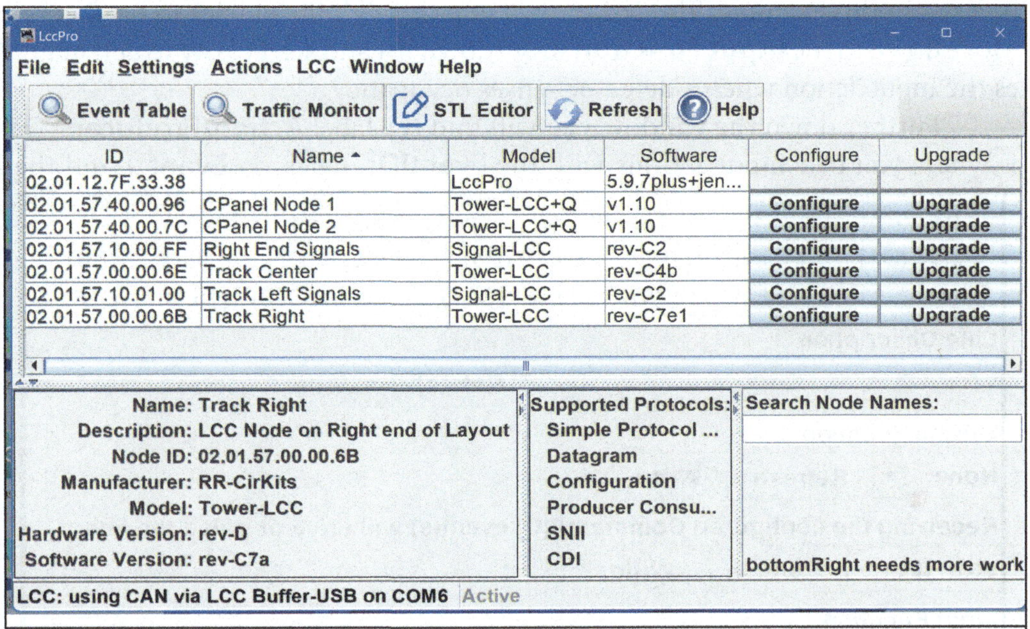

**Figure 6.6 JMRI Displays A Tree Representing The Lcc Network**

play shows several TowerLCC nodes that need to be updated. The latest release at this time is C7e1, which I installed on the last node just before I recorded this screenshot. I will update all the other TowerLCC nodes before continuing. The two SignalLCC nodes were also updated, because one of them supports the KATO turnouts on the Left ladder. The TowerLCC+Q is already at the latest release as I verified on the RR-CirKits website.

Referring to Figure 6.6, many of the Block Occupancy sensors are on the Center Track Node. The yard tracks will be indicated on the control panel. Picking the second track from the top (track 4) i see that the Producer is in the Center Track Node, Line 11 (referenced by "C11" meaning Center 11) and the indicator LED is controlled by node 1, line 16 on the control panel. That tells me where to go for configuration. Configuration will use the existing Evens to ensure compatibility with the unchanged parts of the layout. I take the EventIDs from the Center Track Node to control the LED on the panel. Select the Configure button in LccPro to open the Configuration for The Center Track Node,

The first section of the Configuration Data that was entered was the Name and Description. Since these were already filled in from the original, I have not shown them here. Once you have entered the name and description, you can hide the lines in the window. Enter whatever you use to keep track of the Nodes, and click on the "Write" button next to these fields. I used position descriptive Node Names — you may want to name your Nodes based on geographical features, track mile posts, or any other description you like.

Opening the Port I/O Segment, the top of the Configuration Data should look like the top of Figure 6.7, I have cut out some of the sections that will not be used in this example. The red arrows step through the entries to be checked. Move down to the **Input Function** drop-down list and select **Low** which matches the input action when a detector senses occupancy.

Further down the window you will find the label **Action/Producers** This is where you pair input actions with the **EventIDs**. Click on **Event 1** and then

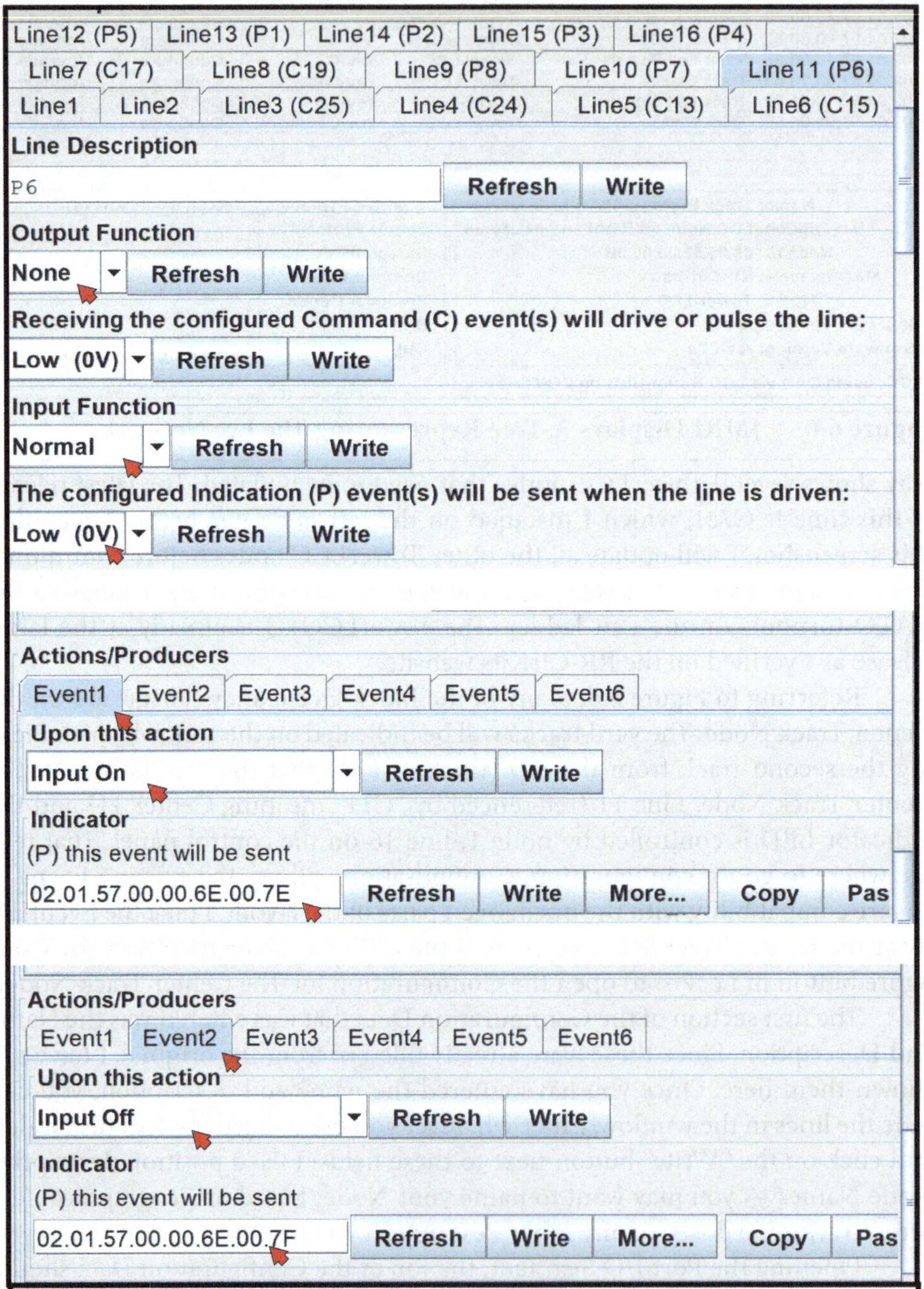


**Figure 6.7 Configuring the Producer for Block Occupancy**

### *Event Usage Listing*

*Below the Event ID there is a list of additional uses of this Event ID. Four lines are listed, but there are only two other uses of this Event. Looking a little closer, one of the uses does not have a Line description — Line 3, Command 1. Decoding this and looking at the data, this is an unused line in the Node, but it does have this EventID in it. Similarly, the "Command 2" use of Line 7 is associated with Event 2 and is selected for "No Action." These are just remnants of the development and will not affect the operation. During development, some attempts did not work out, so Event IDs were left in these lines.*

choose **Input On** which means that when the input becomes low (ground) this event will be sent to the network.

Configuring the second Event is simply selecting **Event 2**, and setting the action to be **Input Off**. Note that the Event IDs are shown as **02.01.57.00.00.6E.00.7E** and the second Event ID is **02.01.57.00.00.6E.00.7F**, which is just the next available EventID. To get these Event IDs to the Consum-

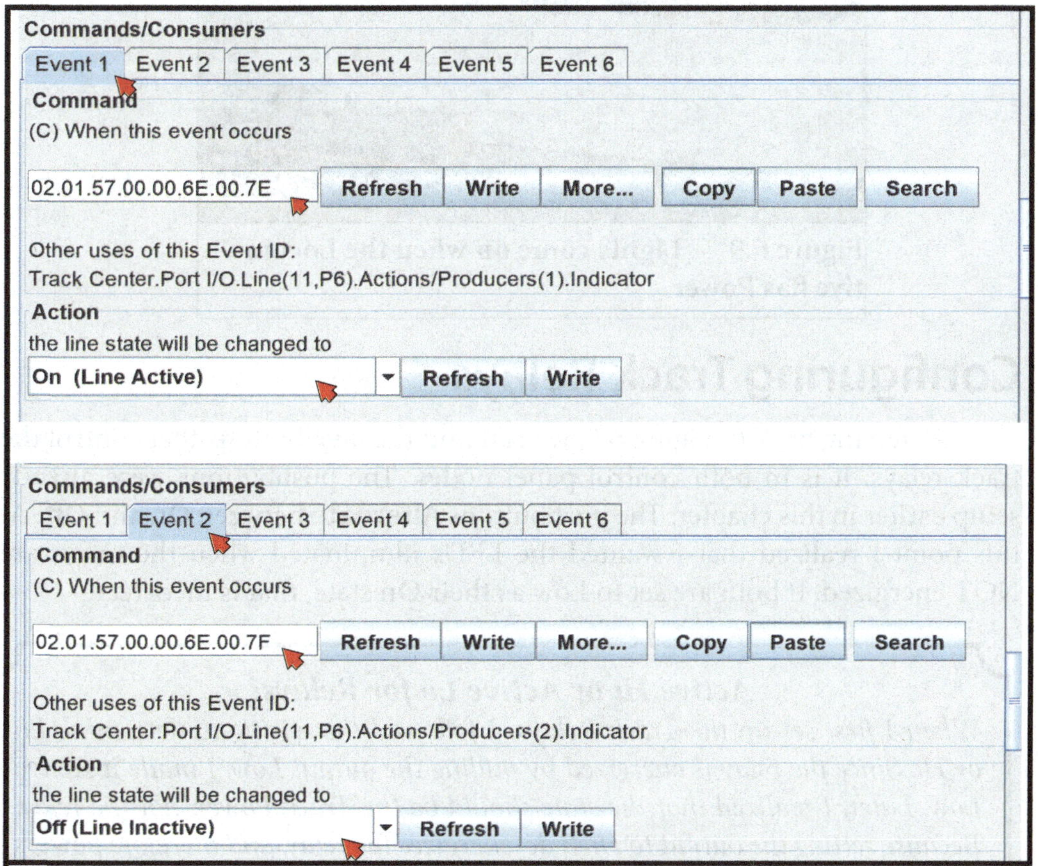


**Figure 6.8 Paste the Event into the Control Panel**

ers, select the **Copy** button. This saves the Event ID so you don't have to type in the numbers when you are configuring the Consumers.

Returning to the JMRI LCC Tree list, you can now open the data for the **Control Panel Node 1** and see that both Nodes appear to be "open" in this window. JMRI allows multiple Nodes to open their Configuration Data simultaneously, so you can jump back and forth between Nodes without Closing them.

Open the Configuration dialog for the Control Panel Node 1. The steps indicated by the red arrows in Figure 6.8 are similar to the Producer. Since this is a Consumer, the Output Function is set to **Steady/Low**, because that is the way the LEDs are wired. The EventIDs are available to **Paste** because I used **Copy** in the Producer side. If you check the two figures, you will see that the **On EventIDs** are identical and the **Off EventIDs** are also identical.

Does it work? In Figure 6.9 the locomotive sitting on the track has a lighted headlight, and the blue LED on the control panel is also lit up. If I take off the locomotive, the blue LED goes out. Now all I have to do is the same steps 11 more times to setup all the block detectors.

**Figure 6.9 Lights come on when the Locomotive has Power**

# Configuring Track Relays

Referring back to Figure 6.5, we can find the pushbuttons that control the track relays. It is in both control panel nodes. The pushbuttons were already setup earlier in this chapter. The pushbuttons **Alternate** between **On** and **Off**. At this point I realized that I wanted the LEDs illuminated when the relays are NOT energized. If both are set to Low as their On state, that is incorrect.

***Active Hi or Active Lo for Relays?***

*When I first set up the Track Relays, I followed the wiring to determine Lo or Hi. Since the relay is energized by pulling the output Low, I made it Active Low. Later, I realized that the state should be the Track Power, not the relay. Because setting the output to High de-energizes the relay, and therefore powers the track, I set the output to Active Steady Hi. This eliminated the need to set the relay and LED separately.*

There are at least two ways to get around this. My original thought was to make both the relay and the LED Toggle (Change state when an event is received) outputs. I discovered that it was possible to get out of sync. So I have what is actually a simpler method: simply set one to active Low and one to active High. That works, but I had to use both Input On and Input Off events. Then I thought "If I am using both events from the pushbutton, why not use the button On to turn on the LED and at the same time Turn Off the relay?" Eureka I said, that is the simplest method. I have already shown how to turn the LED On and Off. I will leave it up to the reader to configure the relays. I just have to repeat this for the rest of the relays.

# Configuring Route Switching

One of the best things about LCC is the ability to use Events in multiple places without running a wire to each use. Also, the Events can be combined in ways that resemble logic equations to perform complex actions. Route switching is one of those functions that is easy with LCC. In the last chapter we worked out the logic (yes, the Interface Tables are a form of logic) for our route switching. Now let us see how to configure the Nodes for this function.

In the period between the previous book and now, a new feature has been added to the TowerLCC and SignalLCC Lines. This feature, called Cascade is a new option in the Producer section. Simply put, if a Line is commanded to change by an Event, an event can be generated to tell another node that a change was requested. To see how this can work, consider Turnouts Ta and Td in Figure 6.5. Td determines which track is used, as long as Ta is set to the Normal state. If the pushbuttons that select the direction for Td are pressed, that turnout will be commanded to change. That command will then generate an Event requesting Ta to change to the Normal direction, thereby setting up all the necessary turnouts.

Looking at a slightly more complex situation,, consider a pushbutton pressed to change turnout Tc. Tc sends an event to Tb to change to Normal direction, and then Tb sends an event to Ta to change to a Diverging direction. If you have more turnouts, just keep Cascading through all the turnouts until you have the entire Route set up.

There is one slight complication on Turnouts Ta and Td. These are KATO turnouts that require different events for Normal and Diverging. (Details about this design are contained in Appendix B.) Both Consumers will need to send the Cascade Event to the Ta Normal direction. Turnout Tb has only one Cascade event which is sent to the Diverging Line for Ta.

The right Track Node has the turnout controls for the Routes that have been configured for the LEDs. The first two turnout motors are Stall Motors controlled by the SMD-8 daughter boards. These are configured as simply "On" or “Off” as seen in Figure 6.12 on page 105. The determination of “Off” or “On” depends on the wiring.

Cconfiguring the single-coil solenoid machines is a bit different because they require two Lines in the Node. In Figure 6.10, the Lines have an additional

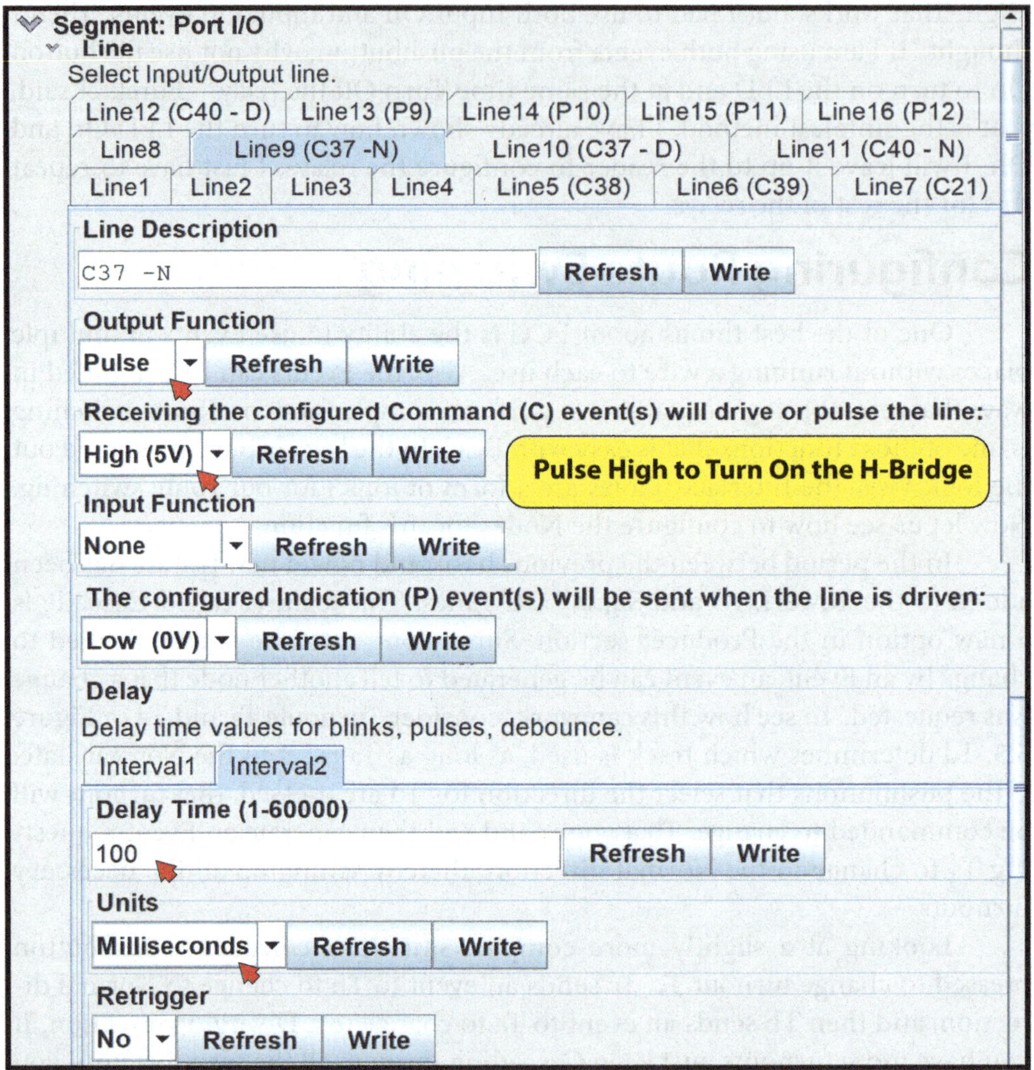


**Figure 6.10 Configuring the KATO Single Coil Solenoid Turnout Actuator with an H-Bridge**

description item — either an “N” for Normal or a “D” for Diverging. I added those designations to ease the configuration after I tested the wiring. The other difference in these Consumers is that they are Pulsed. As before, that means that the signal only stays on for the length of time specified, in this case, 10 milliseconds. That number was determined empirically by testing. Longer would not hurt and would ensure a complete closure of the turnout. Too long will over heat the coil, so limit the time to two or three times the minimum that reliably throws the turnout.

To show the Cascade feature, I went down to the Producer section. Usually we think of Producers as Inputs, but in this case it creates an event without an electrical input. The feature is started by selecting Cascade as one of the options

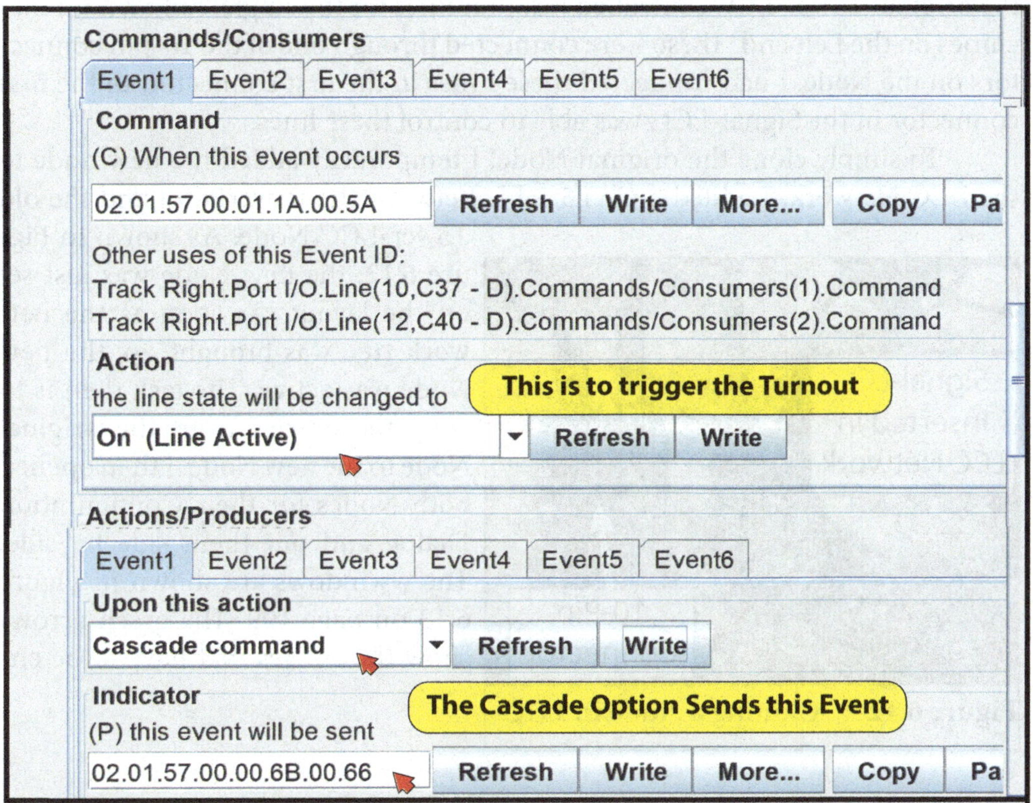


**Figure 6.11 Setting up a Cascade is simply Obtaining an Event to use**

in the **Upon this Action** pull-down. Then it is just a Copy and Paste to use it.

As usual, the rest of the routes are configured using the same steps. With the Annotated Control Panel drawing in hand, the correct settings can be entered. Testing as you go will make sure you are getting the correct settings. Creating the Producers first makes testing easy. If you still have unexplained results, check the JMRI Traffic Monitor to see the Event traffic. In my case, I found that I was sending Events when I did not want to. Seeing the Events on the Traffic Monitor helped to find the bad entries in the Configuration Data.

# Cloning Nodes for New Capabilities

If some new capability is desired, additional Nodes may be inserted into the network. Sometimes these Nodes may take some of the functions out of the existing Nodes. To show how this is accomplished, I have included this section from the original book, unmodified. In the Demonstration Layout, I need to get ready to implement the next chapter on signaling, so I wanted to enhance the Nodes on the Track Board. Looking at the Left end, I knew that the Left Node was very lightly used, so I decided to replace it with a Signal-LCC Node to gain the LED outputs on that Node. This is a very simple swap for the wiring, since the original Tower-LCC controlled just four lines for the single-coil turnout machines on the Left end. These were connected through one of the 10-pin connectors on the Node. I had connected these Lines in the first connector, so the first connector in the Signal-LCC was able to control these lines.

To simply clone the original Node, I temporarily added the new node to the LCC Network. I added a CAT5 cable between the PowerPoint and the old Tower-LCC Node. As shown in Figure 6.12, the new Node was just set on the layout. As soon as the network tree was brought up, the new Node was active. The task then is to copy the entries from the original Node to the new Node. I then opened both Nodes for their Configuration Dialog and put them side by side. The wwindows are shown in Figure 6.13 on page 109. The green arrows show the entries that have to be entered in the new Node. The Line tab is the first entry and then the top data can be entered. Working down from the top, the Event number tab brings up the Event ID **02.01.57.00.00.1E.00.5A**. Copy this by selecting the "Copy" button on the original Node, then select the "Paste" button on the new Node. Continue to select the change from the "Line Status will be changed" drop-down list. Finally, at the bottom, select the "Save Changed" button to save everything to the Node.

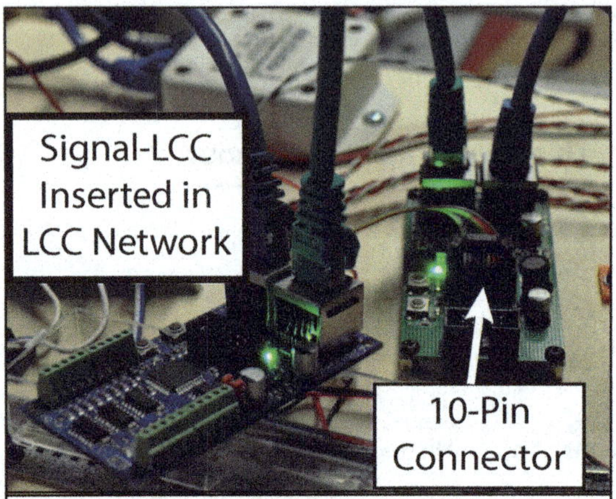

**Figure 6.12 Cloning a Node in LCC**

Notice that this figure is cloning Event 2. That just happened to be the event tab that was open at the time this window was captured. It really doesn't matter what order you clone the entries. The important thing is to select all the Lines to be cloned and then select ail Events that perform an action. If the action is" None" there is no reason to copy the Event ID. Events are easier than Lines, because there is less to copy and the Event ID is copied by the "Copy" button. This is shown in Figure 6.14 on page 110.

After all the Lines and Events are copied, the Node should be ready to replace the original Node. To check that, I moved the 10-pin plug from the original Node to the new Node and tested all the routes. Every one worked correctly, so I replaced the original Node. Figure 6.15 shows the Node installed on the layout, ready to add the wires for Signal heads.

### *Wiring or Configuration?*

*In these turnout configurations, the configuration has been determined by test. For example, the Stall Motor machines are both "Off" even thought the Interface Table clearly shows that they need to be different. Also, I noted that both the specific Line and the pulse length were determined after the wiring was complete. This is one of the great things about LCC, it can adapt to the wiring by changing the configuration.*

Cloning a Node is relatively simple, although it is repetitious. The ability to have two windows open at the same time helps. The process shown here works for either different Node types or the same types. If the Node is a replacement, The best way to clone the Node is to use the Backup/Restore buttons on

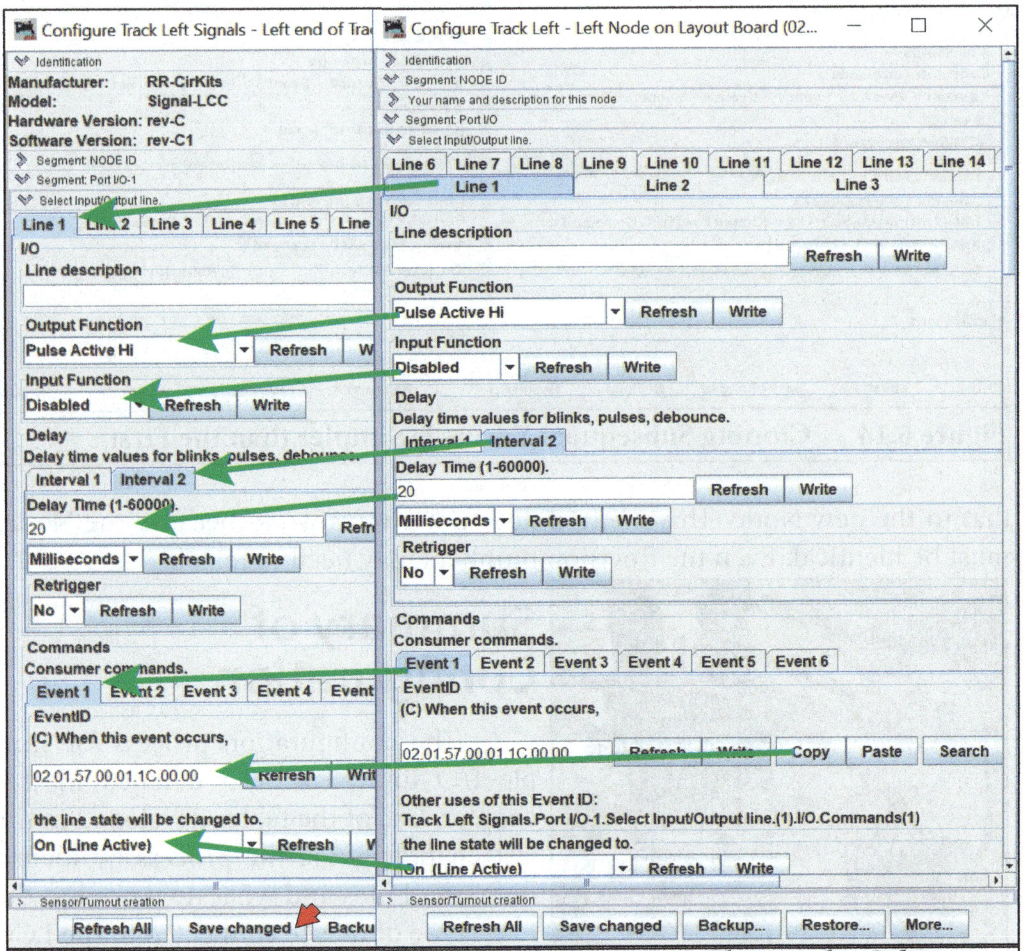

**Figure 6.13 Cloning a Node Copies from one Window to the Other**

the bottom of the window. This makes a full copy of the Node, and then restores

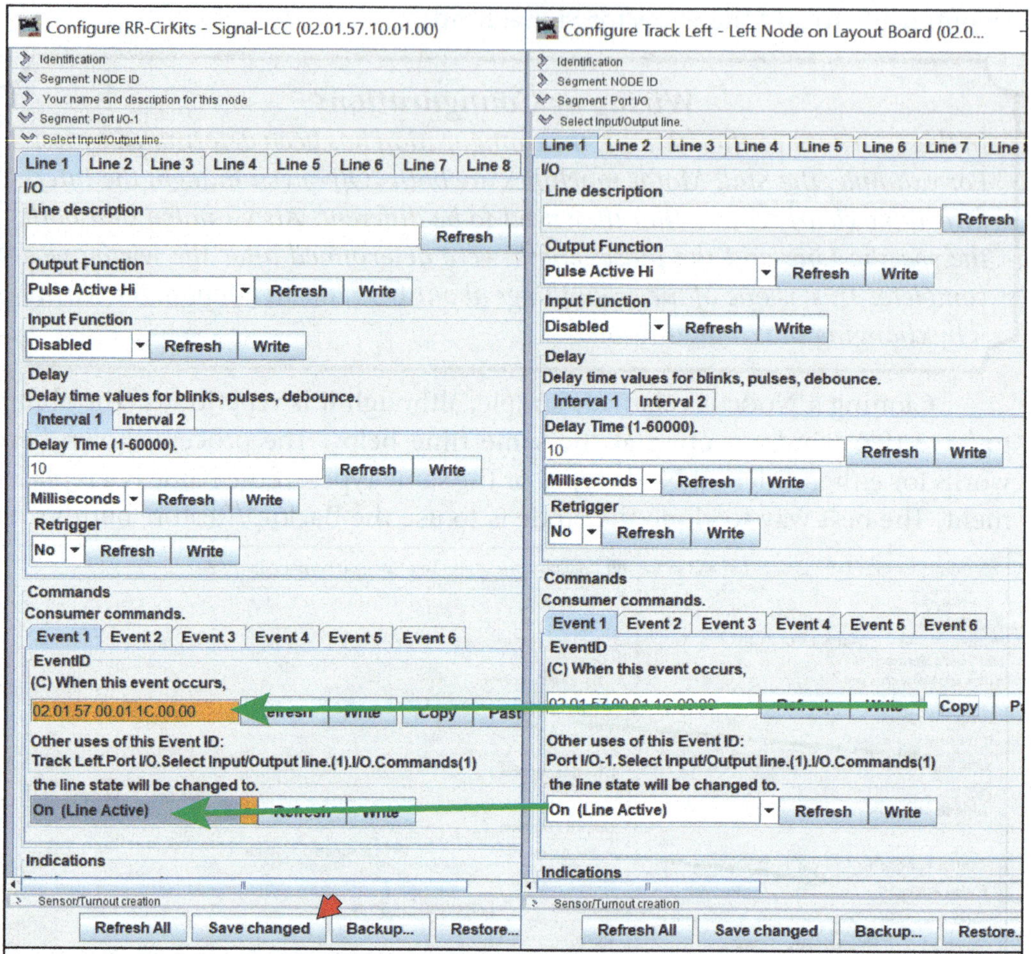


**Figure 6.14 Cloning Subsequent Events is Simpler than the First**

that to the new Node. This method does not always work, because the Nodes must be identical. Even the Revision numbers may need to match.

**Figure 6.15 Node is Replaced**

## Summary of Configuration

The configuration process for simple I/O lines has been detailed in this chapter. Use of the tables developed in the last chapter guide the process by identifying the Lines and what they need to do. Each type of device connected to the LCC Nodes has its own needs in configuration. Some inputs are continuous, some are momentary, but the options allow either. Outputs are more varied, since the outputs can also be pulsed or toggled as well as continuous.

Starting with the Producers allows testing as you go. Event IDs are obtained from the Producers and copied to the Consumers. There is seldom a need to type in the entire Event ID, since it can be copied from the Producer to the Consumer.

Flexibility in configuration allows the Node configuration to change based on the wiring, rather than changing the wiring to match the configuration. The ability to perform configuration without climbing under the layout makes this very feasible. If the wiring had to be changed, several expeditions under the layout would be necessary, including some with a hot soldering iron. This demonstrates a large advantage for LCC.

# Chapter 7 – Logic and Signaling

One of the most visible applications of LCC is signaling. Every layout sparkles when lights are present, and signals are dynamic lights that catch the eye. Unlike the prototype, model railroads usually have short blocks where the conditions in the next block are easily seen. In some cases, signaling can aid the operator by providing information. Regardless of the reason, eye candy or practical information, signaling is desired by many modelers. The goal of this chapter is to introduce the development of signaling using LCC.

Many techniques have been used to implement signaling on model railroads. Prior to LCC, dedicated software or hardware was used to provide signaling. In most cases, a general-purpose computer was used to provide the processing necessary to determine the signal appearance. JMRI is one of those systems. Some modelers have implemented signaling using JMRI as a driver for the LCC Nodes. This may be attractive in that it reduces the LCC Nodes to be just lamp drivers (or is it LED drivers?) This solution does not exploit the full capability of LCC, because it requires a full computer to run all the signals. In this chapter, I will not use JMRI beyond the previous level — use JMRI to configure the Node.

To implement signaling, a few basics will be needed. These are in understanding the logic operators that are necessary to determine the correct indications for a signal. After understanding the logic operators, I will define some signaling terms and then proceed to step through a small implementation based on the demonstration layout built-up in previous chapters. To avoid excess pages and excess boredom, I will only implement a single direction on a portion of the layout.

# Logic Operators and Boolean Variables

Binary arithmetic is everywhere. From Smart phones to toasters, everything seems to have micro-computers. The representation of numbers inside these devices is called binary, which means that it is one of two states, 0/1, "Yes"/"No", or "True"/"False." The name "Boolean" is a tribute to George Boole, an English mathematician that originated the methods in the mid-1800s. He provided the first definitions of the operators that we use today. The basic operators are shown in Figure 7.1. In words, the operators are:

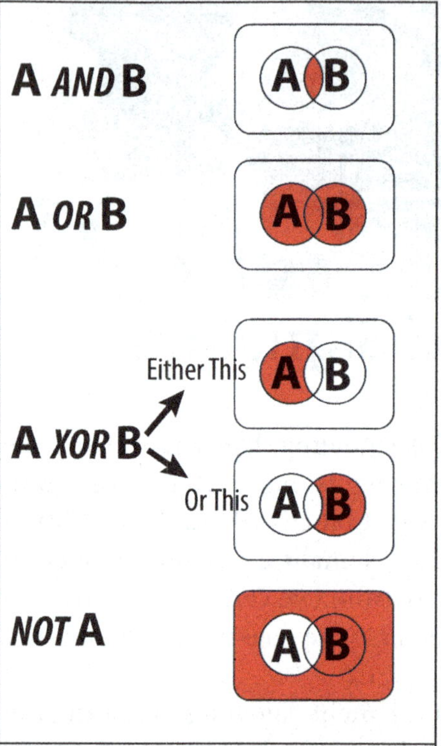


**Figure 7.1 Logic Operators for Signaling – Red indicates when the Relation is True**

- A *AND* B — True if both A and B are True; This is a common operation in routing where turnout A and turnout B have to be in the correct position to get to a ladder track.
- A *OR* B — True if either A or B are True; In signaling, if the next block is occupied, or if the next switch is set against you, then you must light the Stop signal.
- A *XOR* B — True only when either A or B is True; A turnout is a form of exclusive or, because it is either one track or the other track, but never both.
- *NOT* A — true if the input is false; This inverts the meaning of the input; for example, Not Occupied is false if the block is occupied.

Modifications of these operators are present in the RR-Cirkits Nodes, These functions are useful in signaling, and are designed to provide either a default "True" or a time-related function. The implemented additions are:

- A *AND THEN* B — detects the sequence of A followed by B. This is useful to determine the direction of travel. Sequential occupancy detectors would become True, and that would allow a detection of the direction of travel.
- *NULL* — No variable is needed for this one. The result is always True. This could be used as a default condition to force a signal to be dark, or force a flashing to indicate an error.

These simple operators can be combined into complex equations to perform complex decisions. In our use, the operators will be used with only two variables at a time. More complex decisions will require intermediate operators to create intermediate results. In the RR-Cirkits Nodes, the Logic setup is loaded with items that are related to the signaling functions, so we will move on to signaling and then return to logic setup after we know more signaling terminology.

# Terminology and Definitions

To make sure that we are talking about the same things in the same way, several terms need to be defined. Many of these come from the prototype, but a few are unique to the way JMRI and RR-CirKits implement signaling. I have defined these in the way I understand them. Since I am only a System Engineer, not a Signaling Engineer, I may have twisted these a bit to make them understandable.

**Signal:** The object that conveys information to the train engineer at the side of the track or in the cab.

**Appearance:** The physical look of the signal, usually one or more lights or a semaphore arm position. The appearance is the end result of the signal logic which conveys information to the train engineer.

**Signal Head or Signal Arm:** A physical piece of hardware that presents an **Appearance** to an engineer. This may be a single light (called a searchlight), a stack of two or three lights, a set of lights that form a pattern, or a semaphore arm.

**Indication:** In Western railroads, **Indication** is the message that is sent to the train engineer. This often implies a limitation on **speed** to the train engineer. In Eastern railroads, the term **Speed** is used. RR-CirKits uses **Speed** in most of their descriptions.

**Aspects:** The term used to identify a group of **Appearances** that have a specific meaning. For example, the **Aspect** named "Stop: can have various **Appearances** depending on the physical implementation of the signal. Signals have been mechanical (semaphores) or lighted (colors shown in stacks like traffic lights, shown in patterns, or one light that changes colors.) A "Stop" **Aspect** could be a single Red light on the bottom of a stack of three, a single Red light, a set of Red lights in a horizontal pattern, or a horizontal semaphore arm.

**Rules and Rule Books:** Each railroad defines a specific **Rule Book** to standardize the actions of the train crew. **Rule Books** include rules for all actions that occur on the railroad. In our use at this point, the **Rules** enumerate the actions for each **Aspect** and define the meaning of the associated **Indication. Rules** also directly translate to **Appearances.** The **Rule Book** is the final authority on what a signal **Aspect** means.

**Control Point:** The combination of sensors and logic at each signal is called a **Control Point.** It provides the information to set the **Aspects** for a signal. A **Control Point** is a defined location for a Dispatcher or Signal to specify an action to an Engineer.

**Track Circuits and Virtual Track Circuits:** In the prototype the signaling is simplified by each signal reporting the key information to the previous signal;. This communication is sent on the rails when the track is not occupied. The information is essentially the **Indication** or **Speed** of the signal. The method of communicating is called a **Track Circuit**. In the LCC implementation **Track Circuits** are defined in the signal that displays the **Aspect** and are made available to other signals by a **Virtual Track Circuit** which is implemented by events sent to the network.

**Signal Logic:** The logic expressions (as above) that determine the **Rule** to be displayed are defined by the specific track patterns at the site of each signal. Combining several variables into specific **Aspects** determines the most restrictive condition that exists. This is then set as the current **Aspect.**

**Signal Mast:** Just a pole on which signals are mounted, but in this book, a **Signal Mast** is the pole, the **Signal Heads**, the **Track Circuit** and the **Signal Logic** that determines the signal information. This grouping is a convenient way to setup signaling, because you define an entire signal before moving on to the next signal.

**Dark Territory:** Blocks that are not controlled by track signals are designated as **Dark Territory**. These blocks are controlled by other forms movement authorization, such as timetable, track warrants, or direct dispatcher control. In model railroad signaling, **Dark Territory** always exists at the end of the layout or at the end of the signaling that has been installed.

**Automatic Block Signaling (ABS):** is the simplest form of signaling. It really is just "do whatever avoids accidents." **ABS** is best suited for following movements in single or double track main lines. It looks ahead 2 – 4 blocks and determines the allowed train movement (speed) for the current block.

**Absolute Permissive Block (APG):** An extension to **ABS, APB** attempts to prevent accidents (collisions, corn-field meets, etc.) for two-way traffic on single track main lines. Additional details can be found in signaling reference material.

**Centralized Traffic Control (CTC):** Building on the earlier signaling systems, **CTC** adds a centralized controller (dispatcher) to the existing signaling. Commands sent from the central controller are received by the local signal and checked to determine if the action can be performed. Additional details can be found in signaling reference material.

**Tumble-Down:** This term is a result of the **APB** enhanced signaling. To protect single-track movements in both directions, signals throughout the single-track section would be set to prevent entry from one end if a train enters the other end. This action was observed on systems with semaphore heads as a consecutive change from Clear (vertical) to Stop (horizontal). The term **Tumble-Down** describes the visual domino effect of successive semaphore arms dropping from vertical to horizontal. More importantly, it is used as a special indication in the communication between signal masts.

Sorry about the length of these definitions, but the material to follow will be clearer with this common set of terms. Note that not all of these terms are consistent with all prototype railroads, or even with model railroads. The important thing is to establish the rules for your railroad and be consistent, even if there are exceptions to some rules. Let's not be picky, this is supposed to be fun.

## Signal Processing Overview

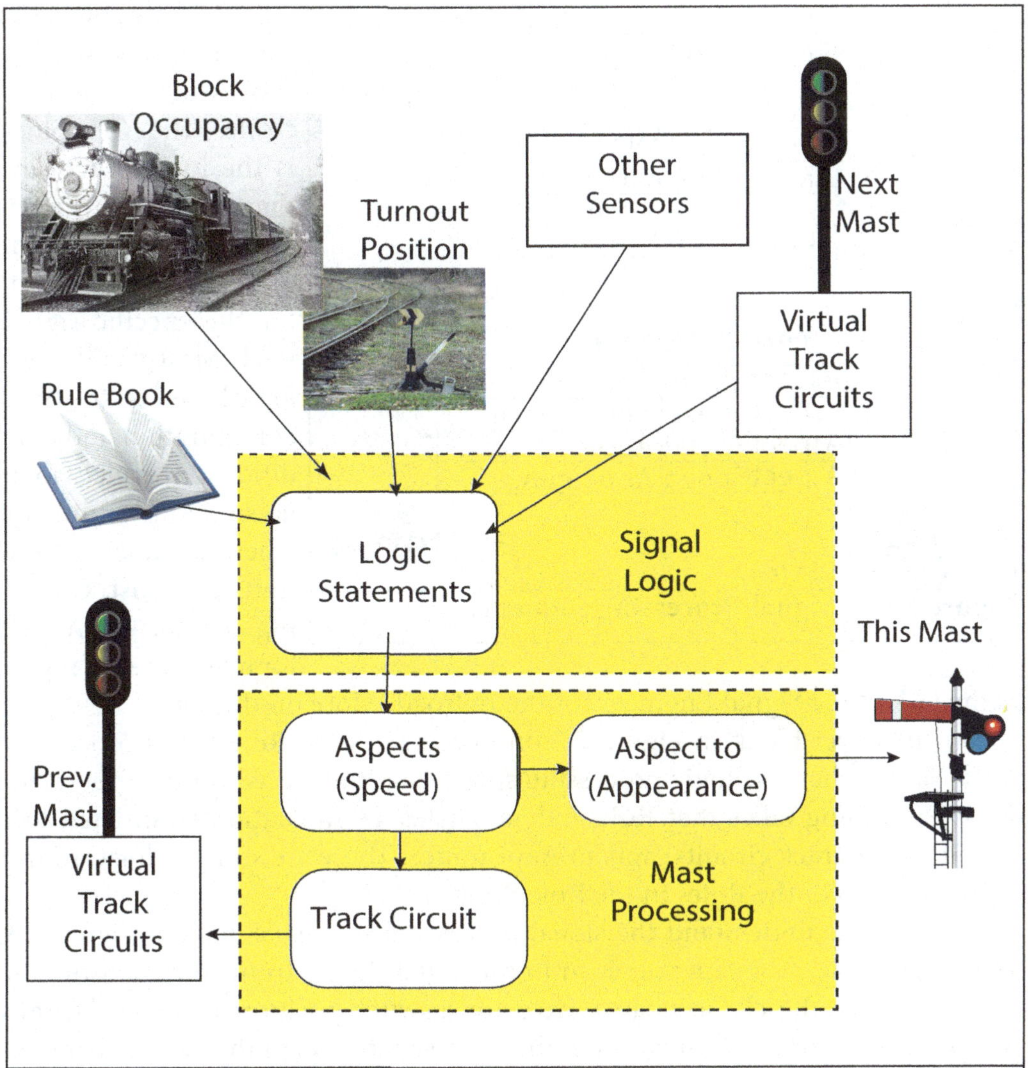


**Figure 7.2 Signaling Involves Several Parts In Both Hardware And Processing**

To understand the steps involved in signaling, Figure 7.2 shows some of the pieces needed for signaling. The basic information needed is at the top of this figure. The primary inputs to signaling are block occupancy, turnout position and the state of the next signal, represented by the Virtual Track Circuits. As we saw earlier, these inputs are specific Events generated by LCC Nodes.

An important source of information for the signaling logic is the Rule Book. Each railroad has a rule book that determines what a signal means. Rule books define the Aspects and the Indication that is intended by the signal. These rules will be used to configure the signaling in the LCC Node.

The outputs on the sides of Figure 7.2 on page 117 are the local signal (mast and heads) and the information sent to the previous signal's Virtual Track Circuit.

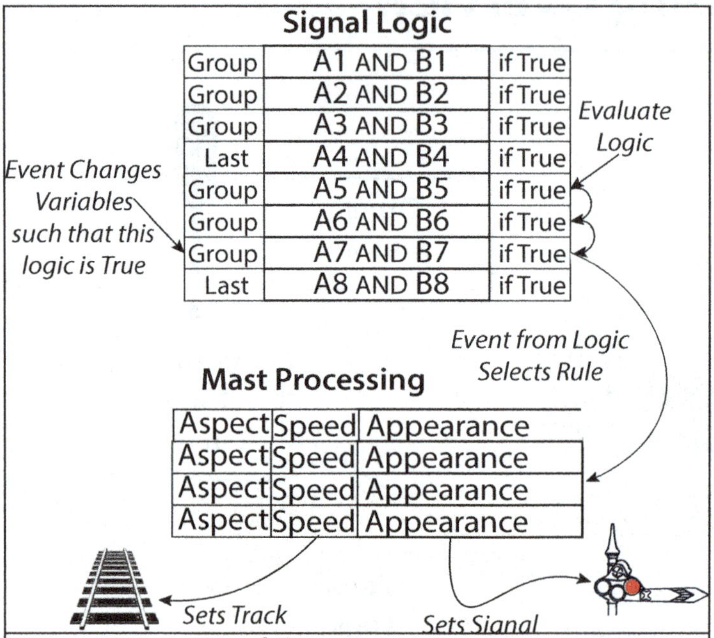


**Figure 7.3 Signal Processing Sequence**

The two blocks in the center represent the processing in the LCC Node. The top block is the logic processing that combines the inputs to determine the correct Rule (Aspect) for the specific signal. Each Mast has its own logic based on the track configuration and conditions. Detailed logic is sometimes difficult when complex conditions can exist. Choosing the correct Aspect for a given set of inputs is the task of the Signal Engineer for the railroad. More on this later.

The lower block in Figure 7.2 on page 117 is the Rule based processing for a Mast. When the logic processing specifies a Rule based on the inputs, the Mast Processing takes that Rule and associates an Indication (translated to a Speed for the track circuits) and an Appearance. These are straight-forward implementations of the Rules in the Rule Book.

To better understand the signal processing sequence, look at Figure 7.3. In this example, an Event comes in to the Signal Logic from some sensor. The sensor may be a Block Occupancy Detector, a switch position, or a Speed from a Virtual Track Circuit. Evaluation of the logic statements in the Group that contains the changed variable starts at the beginning of the Group. The first one to evaluate as True sends an Event to the Rules in the Mast Processing. Notice that the logic statements are grouped by the Group designation and are terminated by a "Last" designation.

The "if True" box on this example indicates that when a logic statement evaluates to True, it generates an Event.

The Event is matched to a Rule in the Mast Processing. The Rule defines the corresponding Aspect, Speed and Appearance. The signal is then changed to the Appearance and the Speed is set in the Track Circuit. The Speed is sent to other users of the Track Circuit, which may cause other Masts to change.

### Things Needed to Define Signals

What did we learn from all of this overview? First, there needs to be a Rule Book, at least in some form. Using a Rule Book from a prototype railroad is probably okay, but it may be too much for a model railroad. Use the prototype Rule Book as a guide. Then the Rules will have to be translated into the appearances that you want. The Rule Book I am going to use is contained in Appendix B. I will only use a part of the rules, because this Rule Book contains situations that will not be needed on the demonstration layout.

In this case, the Rule Book helps to define the location of the signal masts. To understand the signals and their processing logic, they must be placed on the layout. In Figure 7.4 the signal masts are placed, and I have brought forward the Consumer and Producer identification for the relevant devices. I expect to use **Output State Command** options as described in Chapter 3 to determine turnout positions, so the turnout Consumer numbers are shown.

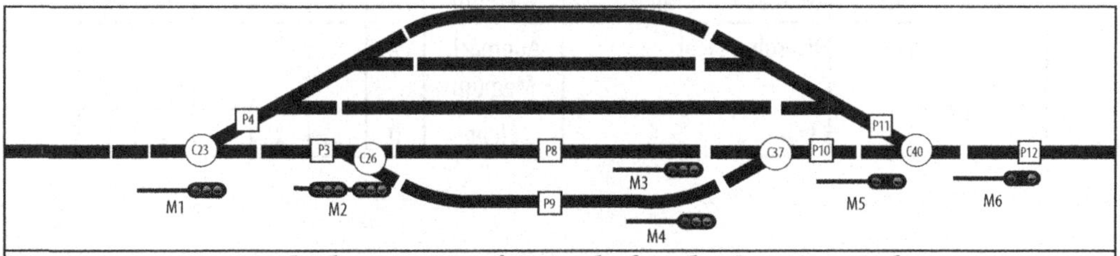


**Figure 7.4 Initial Placement of Signals for the Passing Siding**

# A Strategy for Defining Signal Processing

We have defined the processing needed to implement signals. The major parts are the Mast Processing and the Signal Logic. It may seem strange, but my method is to back into the processing starting with the Mast Processing. I will also work from the right end of the track to the left end. This is because I am going to only implement the signals for a train going from left to right. In Figure 7.4 I have only placed signals on the right-hand side for that reason.

Next, I recognize that the Rules are specific to a Mast, so I will define the Rules needed for each Mast. This will be a help when I get to the logic statements. I created a table to define the Mast entries. I use Excel from Microsoft, but any spreadsheet program or even a piece of paper and a ruler will work. I created columns for the Mast number, Rule number, Aspect Name, Speed, and a set of columns for the colors displayed on the head. Since I am only using color light signals, I don't have to worry about semaphore arm positions. The table fragment is shown in Table 7.1. I have combined several Masts, because they are identical. Masts M1, M3, M4 and M5 all have the same aspects, so they are shown together. Mast M6 is interesting because it has only two Aspects, Stop

**Table 7.1 – Mast Processing Worksheet**

| | | | | Appearance | | | | | |
|---|---|---|---|---|---|---|---|---|---|
| | | | | Upper | | | Lower | | |
| Mast | Rule# | Aspect Name | Speed | G | Y | R | G | Y | R |
| M6 | 54A | Stop | Stop | x | x | 0 | | | |
| | 54F | Approach | Approach | x | 0 | x | | | |
| M5, M4, M3, M1 | 54A | Stop | Stop | x | x | 0 | | | |
| | 54F | Approach | Approach | x | 0 | x | | | |
| | 54I | Clear | Clear | 0 | x | x | | | |
| M2 | 54A | Stop | Stop | x | x | 0 | x | x | 0 |
| | 54C | Approach Restricted | Slow | x | 0 | x | x | x | F |
| | 54F | Approach | Approach | x | 0 | x | x | x | 0 |
| | 54D | Approach Diverging | Medium | x | 0 | x | x | 0 | x |
| | 54H | Diverging Clear | Approach Medium | x | x | 0 | 0 | x | x |
| | 54I | Clear | Clear | 0 | x | x | x | x | 0 |

and Approach. This is because the block that it governs is the end of the layout. Showing a Clear Aspect would allow the engineer to travel at normal speed, which would be a disaster when the train went flying off the end of the world.

Note that no Mast needs more than four Rules. Even though the Rule Book defines nine Rules, Many do not apply to every Mast. This allows the Signal-LCC to limit the number of Rules on a Mast to eight. If you really need more than eight Rules, Masts can be linked together to provide multiples of eight.

The Aspect Name and Speed column options are defined by the RR-CirKits pull-down lists. Since these are arbitrary names, translate them to your railroad. There are 32 different Names and eight Speeds to choose from. The Names are only for reference and do not have any meaning in the processing. However, the Speeds are the entries in the Track Circuit and will be used in the Logic Statements as variables offsets in the Virtual Track Circuits.

The Rules/Aspects defined in the Mast Processing Worksheet are selected by one or more Logic Statements. In this chapter, the Logic Statements are those used in the original TowerLCC and SignalLCC. Use of the current TowerLCC+Q and SignalLCC-32H will be covered in the next chapter. Definition of these Logic Statements requires defining the input variables and logic operator for each statement. With the Rules already grouped by Mast, the Logic Statements can be grouped. Referring back to the Signal Processing in Figure 7.3 on page 118, Processing of Logic Statements starts with the first statement in the Group that contains the variable that has changed. The Groups are probably defined as a Mast, although it is not required. Processing proceeds from the first statement in the Group until an Exit is reached. Usually the Exit is due to a statement that evaluates to True, but an Exit on False is also an option.

**Table 7.2 – Signal Logic Worksheet**

| Mast ID | | Var 1 | T | Function | Var 2 | T | T/F Action | When Out | Where to |
|---|---|---|---|---|---|---|---|---|---|
| M6 | L | P12 | C | V1 Only | | C | T - SX | Immed | M6 - 54A |
| | | | | | | | F - SX | Immed | M6 - 54F |
| M5 | G | !C40 | C | OR | P11 | C | T -- SX | Immed | M5 - 54A |
| | L | M6 - Stop | C | V1 Only | | C | T - SX | Immed | M5 - 54F |
| | | | | | | | F - SX | Immed | M5 - 54I |
| M4 | G | P10 | C | OR | C37 | C | T - SX | Immed | M4 - 54A |
| | L | M5 - Stop | C | V1 Only | | C | T - SX | Immed | M4 - 54F |
| | | | | | | | F - SX | Immed | M4 - 54I |
| M3 | G | P10 | C | OR | !C37 | C | T - SX | Immed | M3-54A |
| | L | M5- Stop | C | V1 Only | | C | T - SX | Immed | M3 - 54F |
| | | | | | | | F - SX | Immed | M3 - 54I |
| M2 | G | P8 | C | AND | P9 | C | T - SX | Immed | M2 - 54A |
| | G | !P9 | C | AND | !C26 | C | T/F - SN | Immed | VX(Siding) |
| | G | !P8 | C | AND | C26 | C | T/F - SN | Immed | VY (Main) |
| | G | M3 - Stop | C | AND | VY | C | T - SX | Immed | M2-54F |
| | G | M4 - Stop | C | AND | VX | C | T - SX | Immed | M2 - 54D |
| | G | M4-Appr | C | AND | !P9 | C | T/F-SN | Immed | VZ |
| | G | VZ | C | AND | C26 | C | T-SX | Immed | M2 - 54C |
| | G | VX | C | V1 Only | | C | T - SX | Immed | M2 - 54H |
| | L | VY | C | V1 Only | | C | T - SX | Immed | M2 - 54I |
| M1 | G | P4 | C | OR | P3 | C | T/F - SN | Immed | VU |
| | G | VU | C | OR | !C23 | C | T - SX | Immed | M1 - 54A |
| | L | M2 - Stop | C | V1 Only | | C | T - SX | Immed | M1 - 54F |
| | | | | | | | F - SX | Immed | M1 - 54I |

Because the statements are evaluated from the first down, it is important to put the most restrictive Rule first and continue with less restrictive Rules. Note that is how the Mast Processing Worksheet grouped Aspects for a Mast. The Mast processing is not necessarily ordered in the processing since the Aspects are set by an Event from the Signal Processing Logic.

Ordering the Signal Logic from most restrictive to least restrictive makes the logic easier in some ways. As the processing steps down through the logic statements, you can ignore some cases that cannot exist because of earlier variable use. In some cases the logic needs to cover all possibilities. If a combination of states exists that has not been evaluated earlier, it may be possible to add it on the end. This would be important in the configuration step, because finding another condition that hasn't been covered and trying to insert it into the logic statements would require re-entering all the statements below the new condition. Adding it to the end would be much easier.

An example Signal Logic Worksheet is shown in Table 7.2. The columns in this sheet have multiple Options. The columns are:

**Mast ID:** Name of the Mast and the Group option which is Group (G), Last (L), or Blocked (B).

**Var1 or Var2:** Source of the Variable. Either a Producer (Including **Output State Command**) or a Track Circuit and Speed. Using the same names as before means that the names are already in the Nodes and identify the Line where that Name is produced or consumed. That makes it easier to find what you want.

**T:** Indicates whether the variable is evaluated only when Changed (C) or when the Variable Matches (M) the current value.

**Func:** Identifies the logic Function to be used. Either ***AND***, ***OR***, ***XOR***, ***AND ⊠CHANGE***, ***OR⊠CHANGE***, ***AND THEN***, ***V1-ONLY***, and ***V2-ONLY***. No invert (***NOT***) operator is provided

### *Comments on Signal Logic*

*Simple signals take very few logic statements. For example, I was able to get three Aspects in only two statements. Of course, there is a lot going on below the surface which is essentially added execution logic.*

*I initially worried that there would be too many logic statements in a single Node. I know that the logic statements do not have to reside in the same Node as the Masts, but still it is less confusing if they do. It seems like I didn't need to worry, because I can compress the logic by using the previous logic statements to eliminate some of the tests.*

*I also worried about the limitation of two variables per statement. By setting up additional internal variables, this is also easier than I thought it might be.*

because the Event can set either True or False. In the worksheet, I use an exclamation point (!) to show inversion — using the False side of the variable.

**T/F Action:** Chooses the action to take when True or False. Actions are SX-Send then Exit, SN-Send then Next, X-eXit, and N-Next. If False is not shown, then the action for False is N-Next. Next means: evaluate the Next statement.

**When Out:** Specifies either Immediate or Delayed. If Delayed, the delay time is set in another entry.

**Where To:** Specifies the destination for the Event generated. This is commonly a Rule in a Mast, but could be Consumer of an Event.

Looking at a simple Mast like M3, the first entry is for Stop. The variables are the occupancy of the next block, indicated by P10, and the turnout position C37. If the block is occupied or if the turnout is against this track (!C37), then we go no further and declare a Stop Aspect. If the first statement is False, the second checks only that Mast M5 is showing a Stop Aspect. The turnout is already known to be aligned, since that was checked in the first logic statement. If the next block indicates a Stop, then this should be Approach. Finally, if neither of the first two are true, then send to the Clear Aspect. This is done by sending an Event from the second statement for the False action.

A more complex example is mast M2. M2 governs the passing siding and has to know more details. The first evaluation is simple — if either of the blocks is occupied, then Stop. For the rest of the Rules, I choose to create an internal state to indicate whether a route — Main line or Siding - was available. I combined the switch position with an empty block to determine which way was open. Variable VX indicates the passing siding is available, and VY indicates the Mainline is available. Using these variables with the state of the far masts, the two approach Aspects can be tested. If neither route is set to Approach, then the last two tests are simply which one to choose to set to Clear.

To expand on using internal state variables, note that the rows in the worksheet that end with an internal variable (which I have designated by a "Where To" entry with a V and a letter) are just logic statements. The difference is that the variables must send an event for both True and False indications so that they can be used as variables in later statements. Also, these logic statements must continue to the next statement. This is indicated by a "T/F-SN" in the "T/F Action" column.

The logic shown is not very complex, although I could have embellished it. I will leave that to the signaling experts out there. As a challenge, I know there are a couple of cases that are not covered by this signal logic. Also, if you look back at the blocks defined previously, you should find a combined block in the M1 Mast. I used another internal variable to combine the P3 and P4 block detectors.

**Table 7.3 – Node Configuration for Signals**

| Node ID | | Description | | | |
|---|---|---|---|---|---|
| | | Line # | Function | Wire Color | Configuration |
| | | **Track Board Left** | | | |
| 02.01.57.10.01.00 | GPIO | 1 | Turnout - KATO - C23 | Brown | Pulsed 100ms forN |
| | | 2 | Turnout - KATO - C23 | Red | Pulsed 100 ms for D |
| | | 3 | Turnout - KATO - C26 | Orange | Pulsed 10ms forN? |
| | | 4 | Turnout - KATO - C26 | Yellow | Pulsed 10 ms for D? |
| | | 5 | | | |
| | | 6 | | | |
| | | 7 | | | |
| | | 8 | | | |
| | Signal LEDs | H1-C | Mast 1 Common | Black | Ground |
| | | H1-G | Mast 1 Green | Green | Out - Active Lo |
| | | H1-Y | Mast 1 Yellow | Yellow | Out - Active Lo |
| | | H1-R | Mast 1 Red | Red | Out - Active Lo |
| | | H1-L | | | |
| | | H2-C | Mast 2U Common | Black | Ground |
| | | H2-G | Mast 2U Green | Green | Out - Active Lo |
| | | H2-Y | Mast 2U Yellow | Yellow | Out - Active Lo |
| | | H2-R | Mast 2U Red | Red | Out - Active Lo |
| | | H2-L | | | |
| | | H3-C | Mast 2L Common | Black | Ground |
| | | H3-G | Mast 2L Green | Green | Out - Active Lo |
| | | H3-Y | Mast 2L Yellow | Yellow | Out - Active Lo |
| | | H3-R | Mast 2L Red | Red | Out - Active Lo |
| | | H3-L | | | |
| | | H4-C | | | |
| | | H4-G | | | |
| | | H4-Y | | | |
| | | H4-R | | | |
| | | H4-L | | | |

The point is that the options in the Signal-LCC can support complex signaling in LCC. These two worksheets can now go to the layout and guide the configuration. In using the worksheets, you must have the Rule Book handy to interpret the Aspect/Rule Names. A designation like 54F may be precise, but the word "Approach" is easier to understand.

# Adding the Devices — Signals and Nodes

As noted in the previous chapter, I discovered that my original LCC network had a Node that was not very busy, both with messages and I/O. The first step in adding Nodes was to replace that Node. I removed a Tower-LCC Node and replaced it with a Signal-LCC node. In the last chapter I showed how I "cloned" the original processing of that Node.

At the other end, I needed to add a Node. I found a good place to add a Node and mounted it like the other Nodes. Adding a short CAT5 cable completed the addition to the network.

Wiring the signal masts is primarily running wires to the Nodes. In the previous assembly, I kept the wiring exposed on the top of the layout, but since the signal wires were through the bottom of the signal, and because the Nodes were on the opposite side of the track, I ran the wires under the layout. Each signal head has four wires to be connected to the Signal-LCC Nodes. I have the screw terminal version of these boards, so I just extended the wires where necessary and inserted them into the terminal strips.

To plan the Node wiring, I modified the Node Configuration table to include the Signal LEDs. As shown in Table 7.3 on page 124, the LEDs are assigned to the terminals as identified in the Signal-LCC manual. These terminals are labeled to make the association with the names in the Configuration and are labeled as Head and Color. This makes it easy to remember the association until you have more than four Heads. An additional point of confusion is that the wire colors are the same for all signal heads, so, you have to be careful in wiring.

Notice that I used the three colors for each Head, although the configuration of the LEDs can use any output for any Mast or Head. The table shown has only three heads, but there are unused outputs for two more heads. This would be configured to use the Hn-L outputs for one of the heads. Added LEDs could be connected to the General Purpose I/O lines like the Tower-LCC. These lines are rated at ±25 ma. which is more than the installed resistors for the signal LEDs will provide. These eight I/Os plus the one leftover H4-L, could control three additional three-color heads for a grand total of eight signal heads.

Using the tables for the two Signal-LCC boards, the signals are mounted on the layout in the assigned locations and wired to the Nodes. I chose to use screw-terminal Signal-LCC boards, so the wires are inserted into the screw terminals on the Node end of the wires. The other end of the four-wire cable is soldered to the wires supplied with the signals. Keeping the wire colors consistent will help.

One consideration is how the signals are wired. There are two possibilities: "Common Anode" or "Common Cathode." These terms are from the days of vacuum tube amplifiers and simply indicate whether the positive leads of the LEDs are connected together (Common Anode) or the negative leads are connected together (Common Cathode.) As I noted above, the signals I have only have four leads, so one of these cases must be true. I tested the signals with a nine volt battery. (There was a resistor already on one lead which was the common lead.) Then, after determining which configuration it was, I was able to check the jumper on the Signal-LCC which selects the configuration.

# Signal Configuration

Configuration will proceed much like the definition. The signals should be configured from the last to the first, the mast and aspects and then the signal logic. This avoids as much back and forth switching as possible. Inevitably, some interconnection is necessary, since the Aspects are set by the results of the logic. Configuring from the last to the first means that the logic that depends on Track Circuits will have the Masts defined when the reference is needed. This should reduce errors. An additional advantage of working from the last to the first is that you can test each Mast as it is entered. If errors are made in an early Mast and you wait to test it until many other Masts have been entered, finding the error may be difficult. The interaction between Masts causes strange results in later Masts. Also, if some condition is not covered, adding logic statements is easier if there are no logic groups below the current Mast.

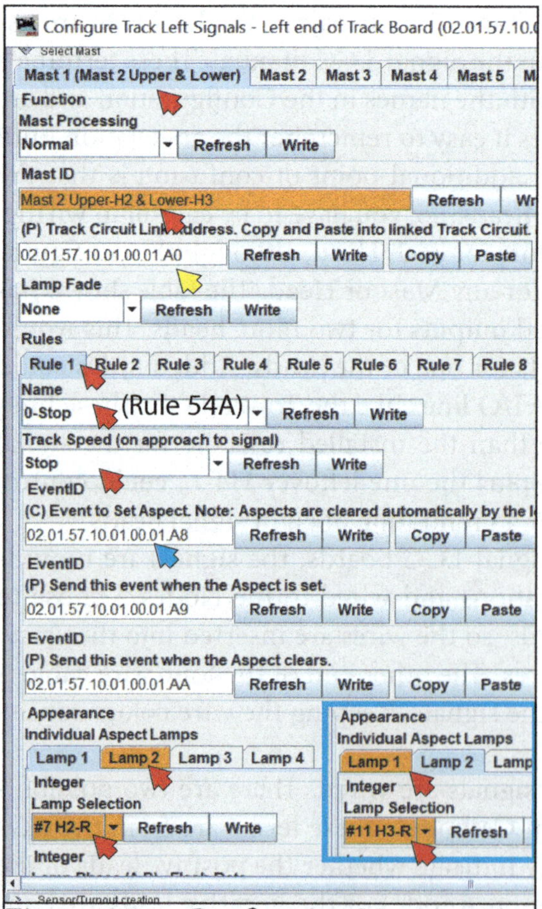


Figure 7.5 Configuring Mast 2-54A

Using the Mast Processing Worksheet in Table 7.1 on page 120, the entries into the configuration data can be entered. Figure 7.5 indicates the fields that need to be set. Going down from the top, I have used Mast 1, because I am working back from the end. These numbers are arbitrary, so I use the next field "Mast ID" to identify this Mast. Note the next arrow is yellow. This field is Read-only and is used to reference the Track Circuit associated with this mast. This will be copied into a track circuit when a Logic Statement needs this Aspect.

Rule 1 is selected for the first rule — the pull-down list has 32 suggestions; none of them are 54A. This field is not used in the processing so pick an arbitrary choice.

Maybe you should also note this on the worksheet for future reference. The next choice is the "Speed" which is used to reference into the Track Circuits. Choose this to match the speed indicated by this aspect. Then, note the Blue Arrow which will be set after the Signal Logic statement provides the EventID for this Aspect.

The last fields to be entered specify the lamps that will be turned on when this aspect is set. Up to four lamps (LEDs) can be set with any aspect. Ordinarily this is to allow multiple heads on a mast. In this case, two heads will light, an upper and a lower. Lamp 1 is selected by its Tab, and the specific output is chosen. Then, in the Blue Box, Lamp 2 is chosen, and its output is selected. The outputs in the pull-down list are the outputs of the Signal-LCC board and are given the same names.

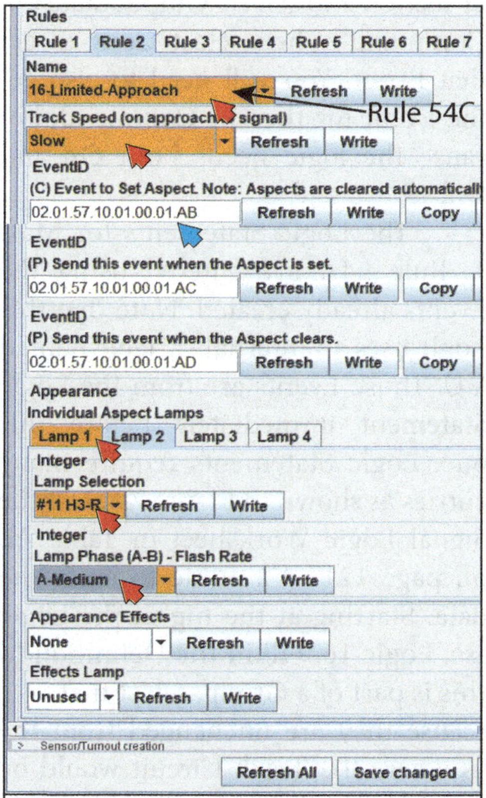


**Figure 7.6 Adding another Rule**

To add another Rule, just go back to the Rule tabs and select the next Rule. In Figure 7.6, I have added the rule 54C. Following the same steps, the Speed is entered, then Lamp 1 is specified for the lower head. In this case, the Aspect requires a Flashing Red, so the Red Lamp is also given a Flash Rate. Two "Flashers" are provided and each has three speeds. This one is set to A-Medium for the initial setting. After seeing the result, this can be changed for aesthetics.

Lamp 2 will be used for the upper head, so it will be set to turn on H2-Y. Since this is a Steady output, no other parameters need to be set.

That's all the Mast Configuration needs to do. The Signal Logic statements will fill in the EventIDs to turn on any Aspect. Notice that the aspects are cleared whenever a new aspect is set, so you don't have to specify what clears an Aspect. When the Logic Group for this Mast evaluates a new Aspect, setting that Aspect will clear the old Aspect before setting the new one.

Some additional fields on this part of the Signal-LCC provide Events when Aspects are Set or Cleared. These may be useful to light an LED on a panel which would then duplicate the signal aspects. Using these Events you do not have to duplicate the Mast Aspect processing.

Enter all of the Mast Aspects using the Rule Book for the railroad. This is just configuring the Masts and Rules like the ones above. Add a Mast, add its Rules, and repeat.

***030445**The next step is to configure the Logic Statements for each Mast. After entering all the masts after Mast 1 and testing them, I will show Mast 1logic to set Aspect 54A. Figure 7.7 shows the first step in this, because this logic needs secondary messages to provide the Events that indicate the turnout position. Since this turnout is controlled by the Lines on this Node, it is easy to find the Consumer C23-Normal and add the Secondary message. Select the tab for Line 1, then scroll down to the Producer section and select "Output State On Command". Then select Copy to get that Event. You will need to also get the Event for the Off Command, because the logic needs both On and Off.

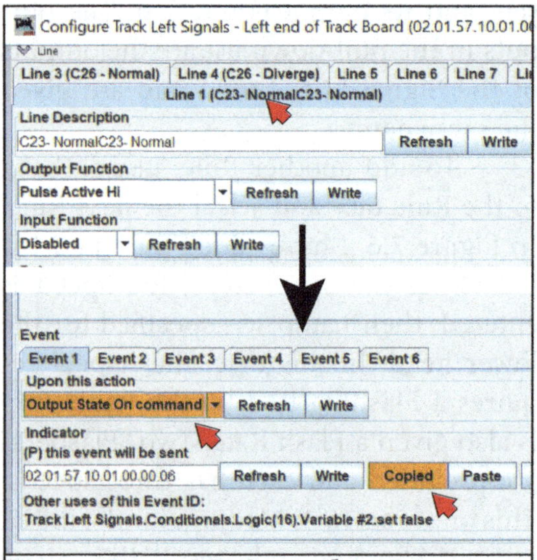


**Figure 7.7 M1 Secondary Messages**

The Logic Statements for Mast 1, Rule 54A are entered using the Events already created. Note that this Logic uses an internal variable named VU. Those Events are from the Logic Statement immediately before this one. Logic Statements require many entries as shown in Figure 7.8. Use the Signal Logic Worksheet in Table 7.2 on page 121 as a guide to enter the data. Starting at the top of the Logic section, select an unused Logic, in this case, Logic 16. Fill in the Definition to identify this Logic Statement, then, since this is part of a Group, select that .

The next two arrows are white, because they are unchanged from the default entries. The second white arrow is where the Track Circuit would be entered. This will be in a later figure. Since this is still the default, this Logic Statement will use Events to set the state of the variable, The next two red arrows indicate the Events that are copied from the previous Logic Statement for variable VU. Select the logic function in the next box. The function was defined in the worksheet. The next few entries specify variable 2 like variable 1.

At the top of the second set of images, the selection for True and False actions are chosen. As specified in the worksheet, if True, Send and Exit. The worksheet implies "Evaluate Next" if False. Both of these are defaults, so the arrows are white.

Scrolling down past the Time Delay section —these apply to the actions to delay sending the events — the next section assigns actions for the True and/or False conditions. Primarily these send an Event, which is then copied with the

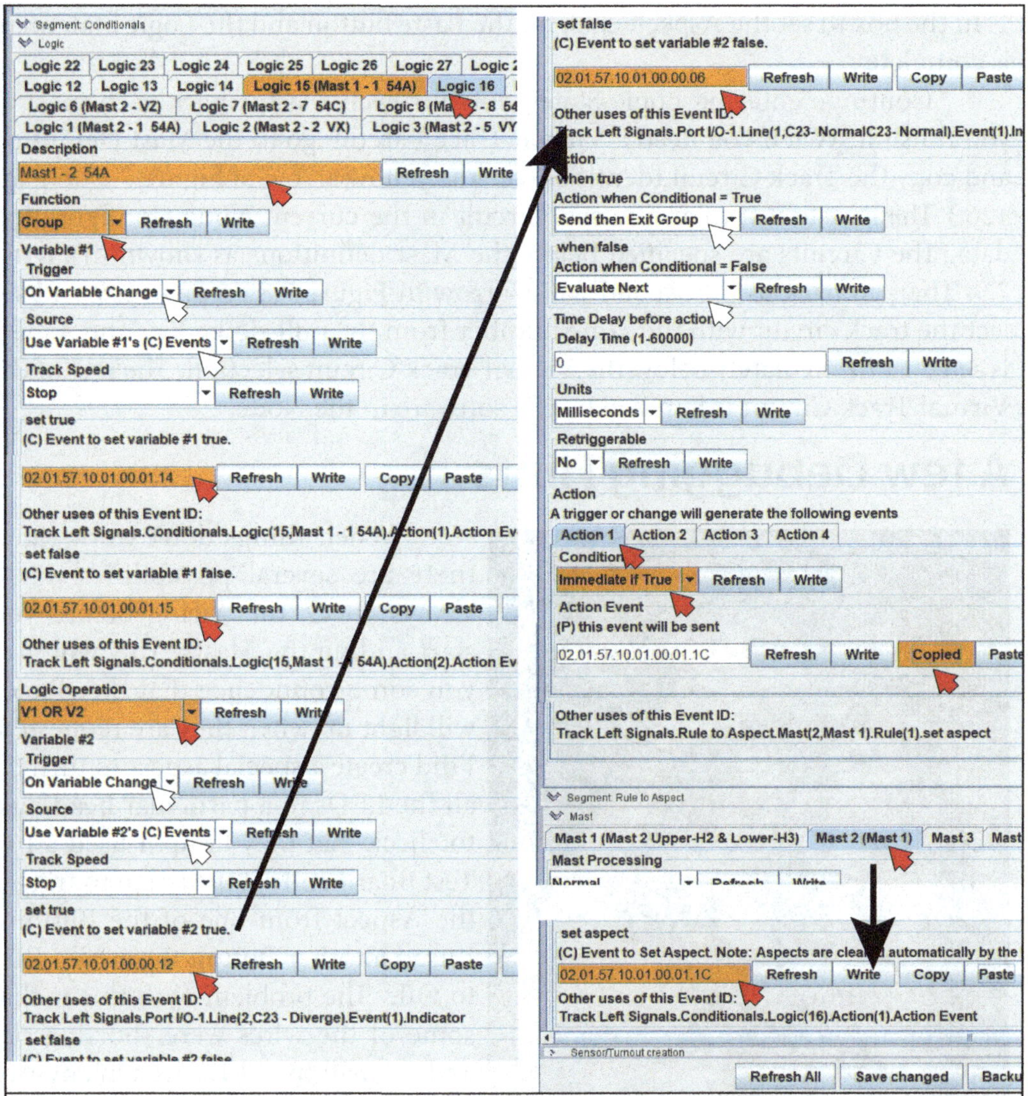


**Figure 7.8 Logic Configuration for Mast 1, Rule 54A**

Mast Logic “Copy” button. This is the Event that will set the 54A aspect, So scroll down to the Mast section, select Mast 2 (which is really Mast 1) and Rule

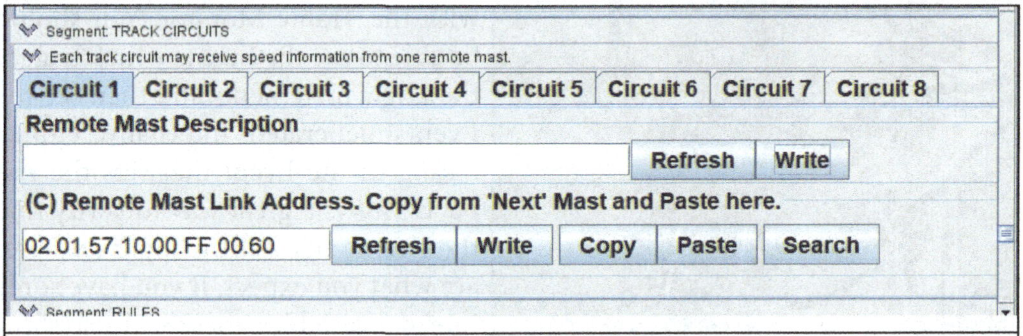


**Figure 7.9 Virtual Track Circuits live Here**

1. In the box to set the Aspect, click on the Paste button and the Logic Statement is complete.

Continue entering Logic Statements for each Mast. Check each mast as you finish it. When you need a Virtual Track Circuit, go to the Mast Definition and copy the Track Circuit Identifier (see the yellow arrow in Figure 7.5 on page 126.) Then paste it into an unused Circuit in the current Node Configuration data. The Circuits are specified below the Mast definitions as shown in Figure 7.9. Then go back to the second white arrow in Figure 7.8 on page 129 and select the track circuit with the same number from the pull-down list. The Aspect is selected immediately below the Virtual Track Circuit selection. The rest of the Virtual Track Circuit is handled by the software in the Node.

# A few Debugging Hints

Figure 7.10 Pretty Signals

When things don't work right, there are several possibilities. First, always check the wiring before you start adding the Masts. This will give you some confidence that the signals will light up when they are requested. I did create a special aspect to turn on all the LEDs in a particular head, just to check that they lit up. I used a I/O Test Board and set the Event to trigger the Aspect from one of the Buttons. I used another Button to set all LEDs to Off. The problem with this is that some of the wires were shorting out and then all the LEDs are On anyway. Next time, I will turn them on one at a time.

Debugging Logic Statements is much easier if you watch the Events with the Traffic Monitor. You should be able to see the Events which trigger the Logic Statements and then see the Events generated internally, culminating in the Event that sets the Aspect. This is a great way to verify that the Events that set the logic variables are what you expect. If you have some internal variables, these will also show up in the sequence.

Missing conditions can be masked by the way the Logic Statements are set. For example, if the last statement has a "False" Event to set the Clear Aspect, then there will always be something displayed. If the Logic Statements are fully defined by "True" outputs, then missing conditions will just fall off the end. To prevent that, I added a Rule to set at the End of the logic. This Rule flashes both Yellow LEDs and is very noticeable, I trigger this Aspect with the False condition at the bottom of the Logic Statements, so there will always be something displayed.

Even after you think you have accounted for all the conditions and you are ready to move on, it may be a good practice to leave a few unused Logic slots for future expansion. If you leave these between Masts, it may avoid the tedious task of moving things out of the way.

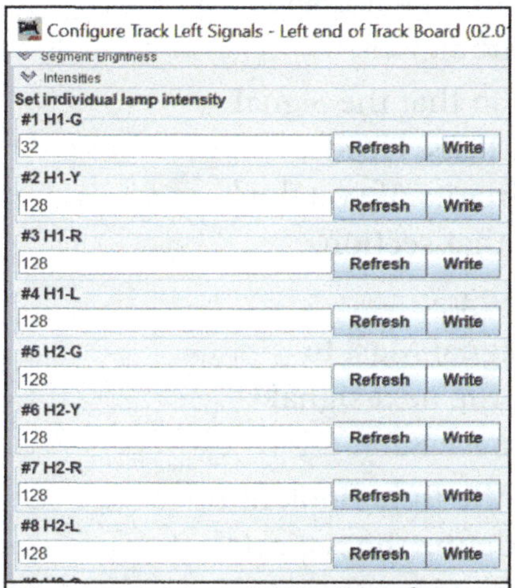


**Figure 7.11 LED Brightness Adjustment**

# LED Brightness

LEDs vary in intensity by color. In other words, a green LED with the same current is usually much brighter than a red LED. Adjusting the brightness of the signal LEDs is a feature of the Signal-LCC Node. The section shown in Figure 7.11allows setting the brightness of the LEDs in a range from 8 to 255. This is accomplished by a Pulse-Width-Modulated waveform generated by the Node.

To understand how these numbers change the brightness of the LED, the first thing to know is that the human eye is a great integrator and does not see the quick on-off pulses shown in Figure 7.12. All the human sees is an average level of brightness. The second point is that the width of the pulses is varied to change the average intensity. The PWM waveforms in Figure 7.12represent the output for settings from 255 to 0, assuming the output is a Common Anode where the LED is dark if the voltage is high.

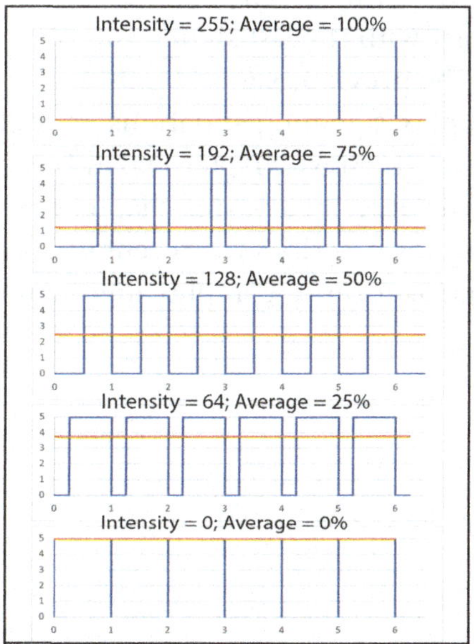


**Figure 7.12 PWM**

Since the light output of an LED is different for different colors and even different LEDs of the same color, it is difficult to predict the changes with the PWM setting. In addition, the human eye is not very sensitive to small changes in intensity. Probably the best way to set these to match is to try different levels until they are satisfactory. I suggest making large changes until you have bracketed the intensity that you want, then try an in-between value until you are satisfied.

## Clear Signals Ahead

LCC signals are very possible using the Signal-LCC board. This chapter has demonstrated the necessary steps to create the signal logic and create realistic signals on your model railroad. The important topics of this chapter are:

- Logic Functions are used in Signal Logic — ***AND, OR, NOT*** are the basic logic functions. These are used in Signal Logic to determine the correct appearance of the signals.
- Signaling needs a common terminology so that the Signal Logic is clear. Significant terms include Masts, Heads, Aspects, Indications, Rules and Track Circuits. If you don't remember the definitions, please review that section.
- A Rule Book defines the Rules for signaling.
- Signals should be defined from the last signal Mast to the first. Since signal Aspects depend on the next signal indication (called Speed) Last to First is necessary.
- A Mast Processing Worksheet translates the Rule Book into settings for the Signal-LCC. Mast Rules define the Aspects/ Appearance and are set by Events from the Signal Logic.
- A Signal Logic Worksheet defines the Logic Statements to enter into the LCC Node. The goal is to define all possible conditions ahead of each signal Mast before attempting to configure the logic.
- Configuration of all the Mast Rules precedes Mast Logic configuration.
- Configuration also works best if you start with the last signal, and test each signal Mast before moving to the previous signal Mast.

These basic capabilities are used in complex signaling situations as well as the simple ones demonstrated in this chapter. The next chapter expands the signaling with current boards. . The initial boards described in this chapter provide a starting point for signaling. The next chapter describes the recent boards which can light up more signal heads and perform more complex logic with greater efficiency.

# Chapter 8 — Signals, Signals, Signals

As layouts grow, LCC becomes both an advantage and a challenge. The LCC network may need additional devices to extend its capacity, budgeting Lines in Nodes may become important and placement of Nodes is also important. The advantages of LCC become apparent when the complexity of wiring is considered. Keeping track of all the wires is a large task for both implementing and troubleshooting. Then, Implementing many signals for large layouts adds many nodes and sensors.

As shown in Figure 7.2 on page <?>, Block Occupancy, also known as Train Detection, is a fundamental part of signaling. In the recent book on signaling, Dave Abeles devoted a chapter to "Train Detection: The Signal System's Foundation." Block Occupancy may be the last thing added before signals. A layout of small to moderate size has little need for block occupancy to operate trains. Only when signaling and dispatcher type operations are desired will the block occupancy become important.

This chapter is written from the viewpoint of a large layout with many blocks and many signals. Efficiency in wiring and network nodes will influence many decisions. The journey starts with a discussion of Block Occupancy methods for large layouts, then discusses the devices to expand the signaling described in the last chapter and completes the signaling with the logic necessary. To demonstrate these things, an emulation will be constructed.

The first subject is on Block Occupancy Detection because LCC needs an efficient way to detect when trains are present in a block. Logic in a Node could provide the same detection as shown, but it would cost several added Nodes. The methods described here are more efficient.

# Block Occupancy

Several methods for detecting trains have evolved over the last few decades. The most recent methods are the result of two things, semiconductor technology and the rise of DCC for train control. Before these developments, train detection was primarily limited to devices like the magnetic reed relay. Today, the most common methods involve optical detection and current detection.

## Optical Detection

Optical Detection depends on either the interruption of a beam of light or reflection of light. Many sensors come as two devices, one for the source and one to receive the light beam. Combined devices are now available to shine on the object being detected and receive reflected light. When using infrared light, these are nearly invisible. Placed between the track rails, they are out of sight and very precise.

In applications like speedometers, optical detectors are perfect, because they are so precise. In Train Detection, the problem is that they are only detecting a very small spot. Since railroad blocks are often longer than the train, detection at a spot does not cover the entire block.

A major advantage of optical detection is that it does not require anything from the object being detected. Trains do not have to install anything to be detected. Optical detection is also immune to interference from other light sources at distances beyond their detection distance, although strong lights can be a problem. Various schemes have been developed to filter out outside interference, so this is becoming less of a problem.

Usually, optical detection includes a delay before the detection is turned off. This carries over the detection state through things like the space between cars and open hopper doors. The delay is often adjustable.

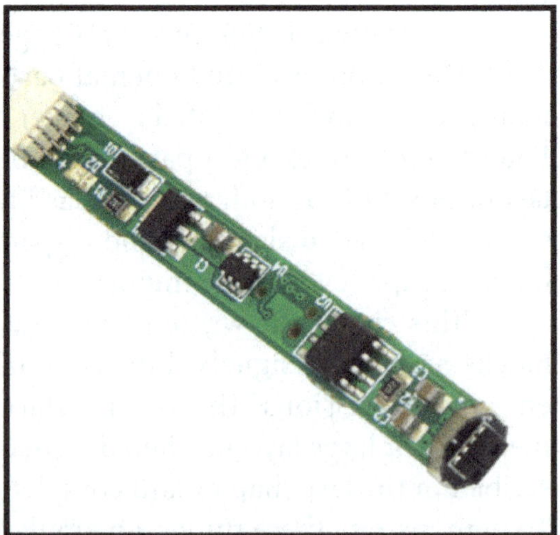

**Figure 8.1 The Iowa Scaled Engineering IR Detector Fits between the Rails in N Scale**

Installing optical detectors has advanced from aiming a beam of light across the track to reflective IR (infrared) sensors. Early optical detectors required a source of light (visible or IR) ands a detector aimed across the track. The source and detector were hard to hide and sensitive to ambient light. Taking advantage of ambient light allowed just the detector mounted between the rails. This solution was okay if there was sufficient light and failed completely during "night operations." It also was vulnerable to anything that was above the track.

Recent developments in IR reflective sensors allow easier mounting ands less sensitivity to other sources. The reflective sensor aims a beam of IR up from the track and aims a detector at the location where the IR reflects off of the car above. These sensors come in either two parts or combined in a single package. Small enough to fit between the rails, the sensing distance depends on the angle of the IR beam intersecting the point where the sensor is aimed. Typically, theses sensors are set at a few inches. If a larger distance is desired, independent source and sensor parts can be used. The IR Sensor in Figure 8.1 is small enough to fit between the rails in N-scale, although it will require cutting out a tie.

### Current Detection

Early forms of Current Detection were used in DC control by detecting the voltage drop across a rectifier. This form of current detection was made popular by Linn Westcott in the 1970s. Applicable to both DC and DCC control, two issues were found in this sensor. First, applying to DC, the sensor only worked when power was applied to the track. In DC control, a stationary train (at a station, for example) does not require current, so no voltage can be detected. This issue is true of all Current detection sensors. DCC does not have this issue because DCC decoders require a small current at all times.

The second issue with the rectifier based current detector is the interaction with the track voltage. The rectifiers cause a small (typically less than one volt) drop in the track voltage. While this is small, it can cause a difference in speed when transitioning from a detected block to and undetected block Connecting to the track also means that the sensor needs added isolation to prevent connecting track blocks accidentally.

With the wide-spread acceptance of DCC in the 1990s, a new form of Current Detection became possible. This detection scheme is based on a device called a current transformer. A current transformer is simply a transformer without one winding. The wire that carries the current to be sensed is passed through the transformer center. In use, a current proportional to the current in the wire is created on the transformer's output windings.

**Figure 8.2 A typical Current Detector from ISE**

In use, the sensor looks like Figure 8.2. The current transformer is on the side of the PCB. You can see the center hole where the wire is to be inserted. One of the two track leads is passed through this hole.

This current detector does not interfere with the tack power. Any voltage drop is very small, only the drop that is caused by the wire resistance. There is complete isolation between the track and the sensor.

The two considerations with this type of detector are (1) it has to be an alternating current like DCC, and (2) there has to be something on the track that requires current.

### Cooperative Detection

To overcome the issues with both Optical Detectors and Current Detectors, a method to combine them has been created. The concept is to use the best features of both Current Detection and Optical Detection to define a block. While the Current detection senses when a train is in the block, Optical detectors can sense when a train is entering or leaving a block. Figure 8.3 shows the movement of a train through a block. At the beginning, none of the three detectors detect anything, so the block is not occupied. As the train moves into the block, first an IR Detector (the red arrow on the Left) comes On, indicating something

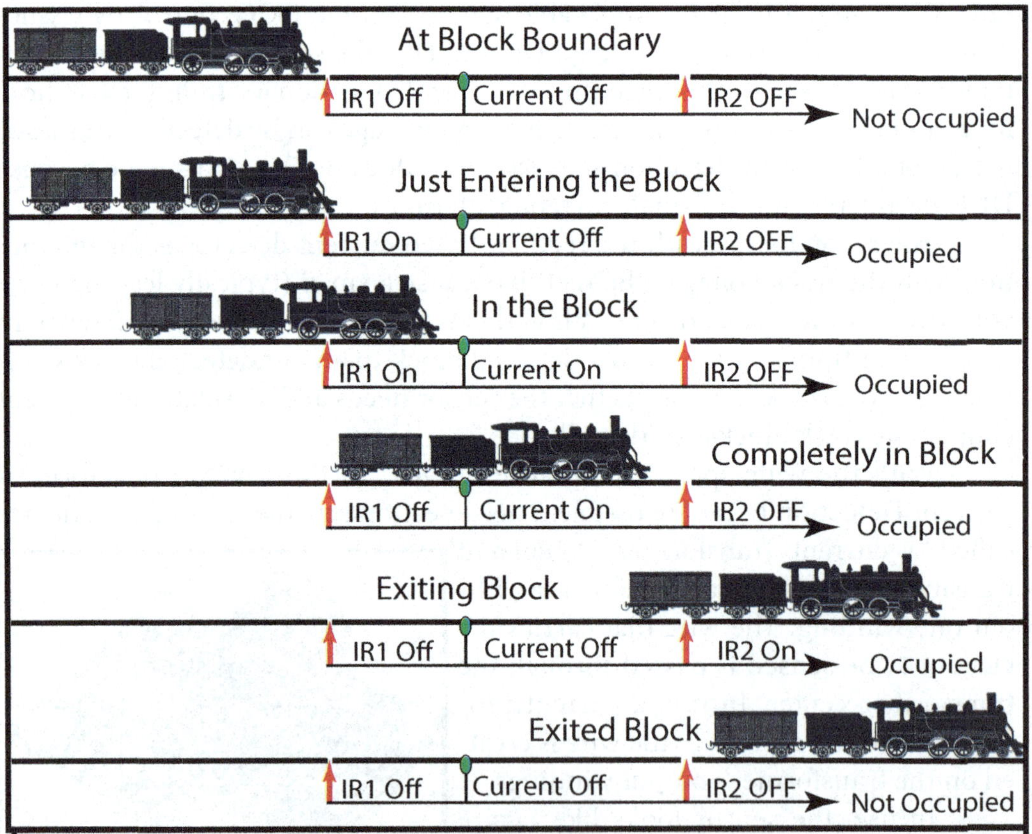


**Figure 8.3 Cooperative Block Occupancy Solves some Issues**

has entered the block, then the Current Detector (the circle around the track) comes On. If the train is shorter than the block, there will be a time when only the Current Detector is On — the block is still occupied. As the train leaves the block, the Current Detector will turn Off, but the Right hand IR detector will stay on until the last car passes over.

Other situations may require additional IR detectors to detect unpowered cars. For example, a siding with a loading operation like a reefer icing platform or any siding that is routinely used to park cars may need additional IR Detectors. The Cooperative Detection uses the wired OR method, so additional IR Detectors can be easily added. Cooperative Occupancy Detection is not a complete cure for all possible conditions. For example, if a train looses a car in the middle of a block, that car may not be detected. The advantages of Cooperative Detection are that it is able to detect trains with cars that do not require power and it is isolated from the track power.

Wiring the Cooperative Detection requires three or four wires to each IR Detector and three to the Current Detector. Two types of Optical Detectors (Triangles and rectangles in red or green) are shown in Figure 8.4. The top detectors

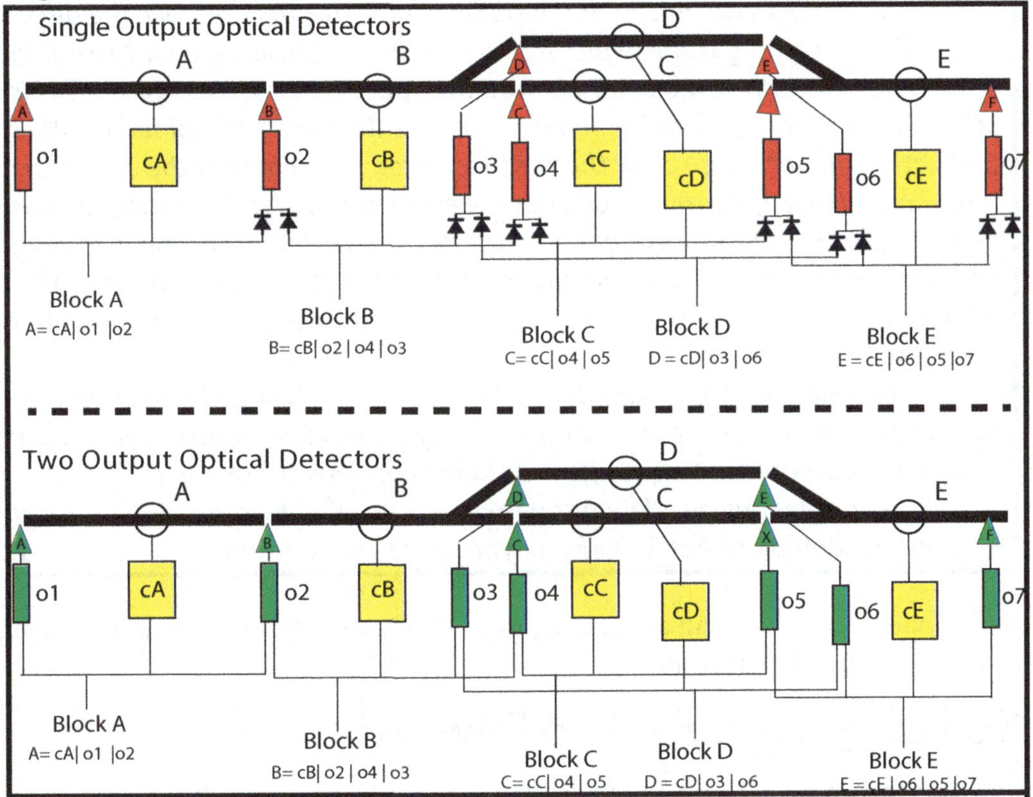


**Figure 8.4 For Cooperative Detection Optical Sensors need two Outputs to connect both Forward and Backward**

have a single output connection and need diodes to keep the blocks separated. If the detectors have two outputs, the bottom wiring is used. In either case, the Current Detectors shown in yellow are wired to provide a logical Or. The resulting logic is shown in the equations below each block. If addition;l IR Detectors are required, simply add them intro the block that contain them. Even with single output detectors no diode is required.

Diodes with low forward voltage are required for reliable operation. Since LCC inputs are low current, low power signal diodes are used. Shot key diodes are preferred because they have low forward voltage.

Combining Cooperative Detection with LCC provides a solution that minimizes wiring and achieves Train Detection for nearly all cases. As noted before, orphan cars are not detected, but that should be rare in most cases, When that happens, the block can always be forced to "Occupied" by moving a car over one of the Optical Detectors.

***Block Boundaries Determine Detector and Signal Mast Placement***

*Deciding on the Bl0ck Boundaries is a task that, combined with the type of signaling, determines where to install the Optical Detectors and the signal masts. My preference is aligned with the CTC positions, even if CTC is not active. The CTC blocks are at the "fouling point" of turn-outs and face the direction of the on-coming traffic. I prefer that position because it is easy to explain to operators that they must stop before they pass a restrictive signal. This is often a spot that is easy to implement because the turnout is joined to other track at that point. To be clear this puts insulated rail joiners at the end of the through and diverging rails. I consider the entire turnout to be in the previous block. Figure 8.4 shows this type of block definition. Signals at the incoming end (the turnout point rails) are placed so that the turnout can be operated without obstruction. This is somewhat further than the block boundary, but keeps the turnout free to move.*

*Intermediate blocks are defined for spacing and speed reduction purposes. These blocks place the signals directly across from each other. These blocks are separated with insulated rail joiners or gaps cut in the rails.*

*In your layout decide what works best for you. If your solution is not strictly prototypical, but it works for you, don't worry about it.*

# So Many Signals, Too Few I/Os

As the photo at the start of this chapter shows, signals multiply with every turnout. Surpassing the capacity of the original SignalLCC is easy with a limit of only four signal heads (five with some mixing of signal outputs) less than two turnouts use all that capability. A simple passing siding would require two SignalLCC boards. Signal processing is inherently slow, so the issue is not the speed of the processor, it is the number of I/Os that can be connected.

The SignalLCC-32H implements a very simple serial connection to provide control for 32 signal heads. The basic unit for this is a small board which contains a integrated circuit to drive three LEDs (Red, Green, Yellow.) These boards have a shift register that connects an input to an output. When the inputs stop, the content of the shift register is latched into the drivers to light the signal head.

The SignalLCC-32H is otherwise the same as the original SignalLCC. It has exactly the same Port I/O and Conditionals. It does not have the Output drivers for LEDs, since it relies on the serial output to drive the signal heads.

Configuring signals for the SignalLCC-32H follows the same process as shown in the previous chapter. It is important to define the signal masts first, using the Rules from the Rule Book. This time, however, the Masts are only there to carry the Heads. The Signal Head data is loaded onto the partial line to be passed to the connected driver circuits. As shown in Figure 8.5, the serial connection can wander around to find the shortest path, as long as you put the proper rules into the Node and select the correct head when displaying an aspect.

After defining the Heads, the steps to define the signals follow the same steps as shown in Chapter 7. You may notice that the number of heads in the

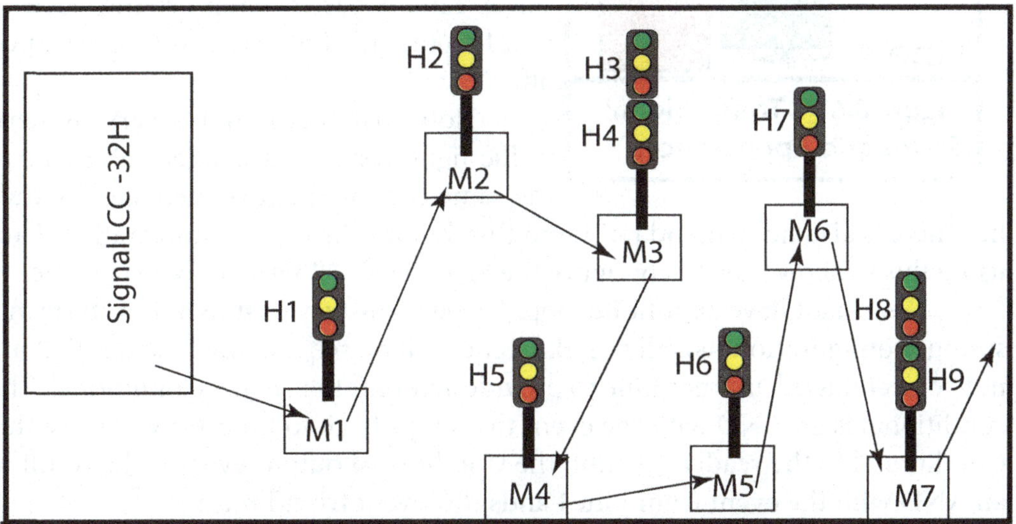


**Figure 8.5 Signal Heads are loaded in numerical order.`**

SignalLCC-32H is exactly the same as the number of Conditionals available. In the next section of this chapter, the STL feature will be described to provide additional logic. In any case, using logic on another node is possible, because the Events provide communication between the Nodes.

## A Simple Demonstration

To test and demonstrate the capability of the SignalLCC-32H, I will design and implement a traffic signal set. The idea is to test the serial connection and to show how the Head data is loaded. No logic will be required for the traffic signals. To show that the serial connection works, I have defined 5 intersections with 4 traffic signals at each one. The states required are shown in Figure 8.6. The starting point is Red for the two signals in the North and South directions. At the same time the East/West signals are Green. This was set when an Event started Line 1. Line 1 will be set as a **Pulse** for a duration of 10 seconds (I am impatient, so I set the times unrealistically.)When the output drops, I will use a Producer Event called "Output Off (Function Lo)" to start Line2. Line 2 will Cause the East/West Heads t change to a Yellow, and it will be a pulse for 2 seconds (Did I hear a screeching of tires and cursing because it is short?) After 2 seconds, the Line2 Producer Event will start Line 3, which also sets the North/South Heads to Green and the East/West heads to Red. Line 3 then does its time and Line 4 starts. Line 4 has to wrap around back to Line1 so that the light runs until it is turned off.

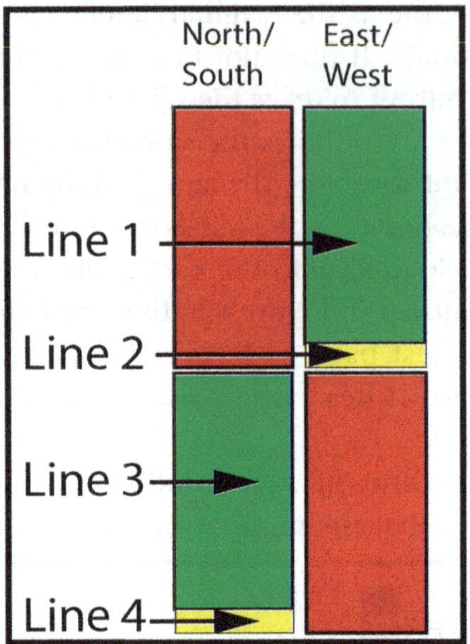


**Figure 8.6 Traffic Signal States and Appearance.``**

Note that this is not a very modern traffic light system—it changes all lights at the same time without sequencing for traffic. There is also no demand built into this, but it is just for demonstration. I am using this to show how to configure the SignalLCC-323H, not to be realistic.

Now that I have an infinite loop, I need some way to stop it. If it starts in a strange configuration like all On, the lights will be sequencing in states that are not correct. I used another Line to provide an event to use in a Conditional. The conditional is an AND with the event that loops back to Line 1. I will leave this Conditional to the reader. [A hint: the Conditional output event is the result of an AND with the event from Line4 ands the event from Line 5.]

This demonstration only implements 4 intersections with 4 traffic signals at each intersection. I have severely compressed the distance between the intersections so that the demonstration is not unwieldy. I also note that only 4 signals per intersection is the minimum for two-way traffic. The demonstration will still be controlling 16 traffic signals with 3 LEDs in each. Therefore, there will be 16 "heads" in the SignalLCC-32H configuration. All heads will use the same ""rules"—Red, Yellow, Green and Dark. Only 5 events will be defined to set all of the signals, the 4 events that set the Lines and one from the On/Off Line. To make this process easier, the signals will alternate, e.g. all odd numbered signals will be Red at the same time all even numbered signals are Green.

Configuring the SignalLCC-32H is all about the Heads, unlike the pre-

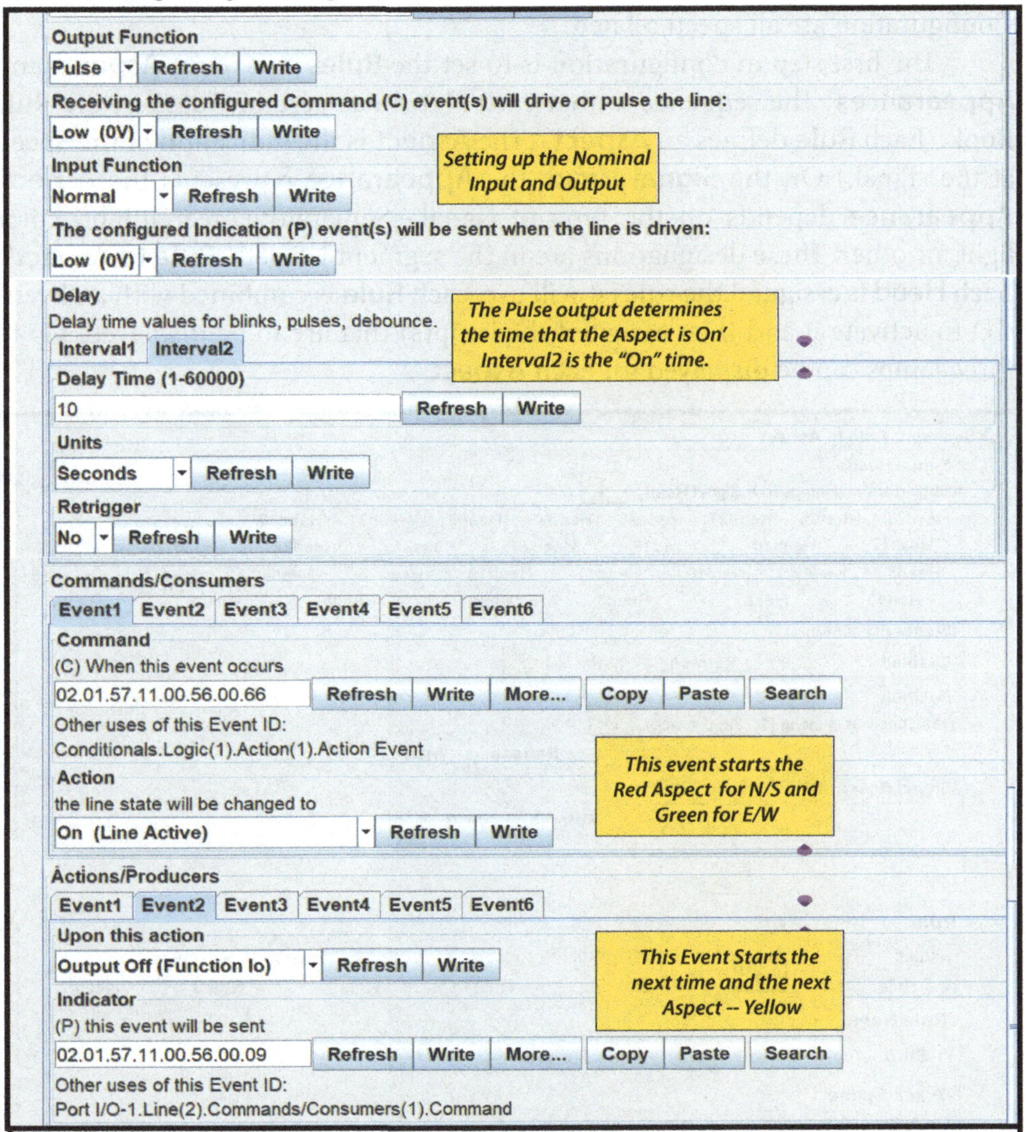


**Figure 8.7 Configuring the Traffic Light Loop**

vious chapter. The Heads are what are specified in the configuration, however there is a way to link together two heads for a single Mast. Many Masts have only one Head, so concentrating on the Heads is nearly identical to concentrating on the Masts. Traffic signals in this example are single head masts.

## Configuring a Single Head Mast

When using RR-CirKits SignalLCC products, remember that the signaling is based on the Eastern railroads and is all about speed. All signal aspects are indicating a speed restriction, regardless of the track routing. Track routing is converted to a speed that is required for the route. This is why the entries in the Configuration are all speed related.

The first step in configuration is to set the **Rules** and their **Aspects** and **Appearances**. The sequence is that each **Head** has certain **Rules** in the Rule Book. Each **Rule** defines an **Aspect**. (The **Aspect** is the indication of the Speed at the signal.) On the **Signal Head**, the **Appearance** represents the **Aspect**. **Appearance** depends on the type of signal—Semaphore, searchlight, color light, or other. These designations are in the segment labeled "**Rule to Aspect**". Each **Head** is assigned the rules it will use, each **Rule** is combined with an EventID to activate it and then assigned the lamp(s) that are to be displayed. Up to three lamps can be displayed for each **Aspect**.

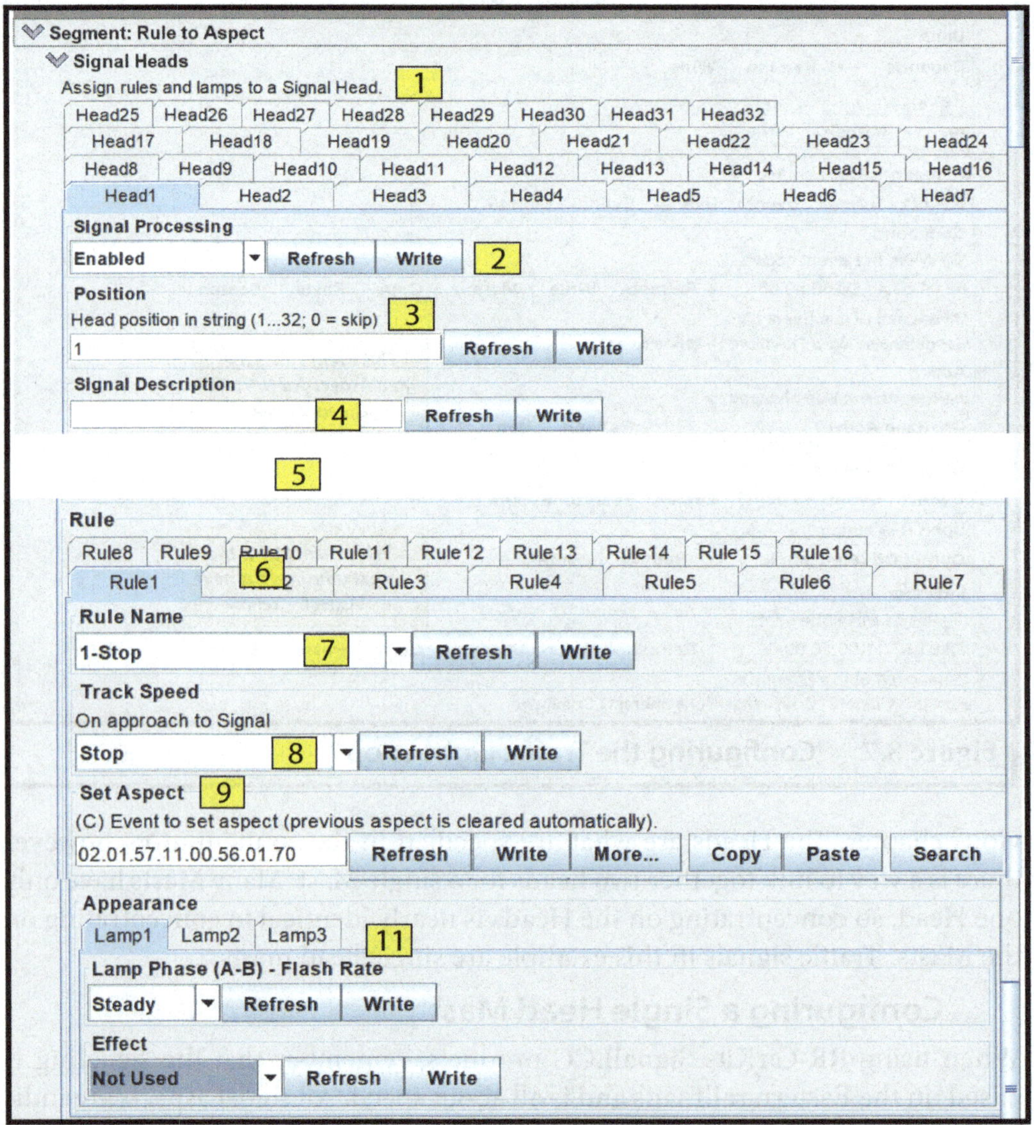


**Figure 8.8 Configuring the Heads using the Rules, Aspects and Appearance. Numbers refer to the Text Numbered List**

Figure 8.8 shows the Configuration fields that need to be filled in. Taking each one in sequence:

1. Select a **Head** from the Tabs at the top.
2. **Enable** in the pull-down list for **Signal Preprocessing**./ This adds the heard to the set that can be sent on the serial interface.
3. **Position** is the physical location on the interface.
4. **Signal Description** is a name for the specific signal.
5. Skip over the fields before the tabs for **Rules**. These will be explained later.
6. Select a **Rule** from the tabs to add to this **Head**. Each **Head** is restricted to 8 Rules.
7. **Rule Name** is selected from the pull-down list. This is restricted to the names in the list so that the **Track Circuit**s can be defined.
8. **Track Speed** is also related to the **Track Circuits**. Pick the most appropriate one for now.
9. **Set Aspect** is where you enter the **EventID** that turns on this **Aspect**. In our case, this is from the Lines that were configured in the section above. When this **Event** occurs, the **Aspect** on this **Head** will be changed to this one.
10. Two Events that are produced by these actions are skipped over.
11. Finally, the **Appearance** of this Aspect is configured. Select the **Lamp** which corresponds to the color you need and then pull-down the poorly named **Lamp Phase**. Fore most cases, the lamp will not flash, or b;ink so search down the pull-down until you find **Steady**. That will turn on the lamp without any blinking.
12. Repeat from step 6 to add **Rules** for this **Head**.

13. Repeat from step 1 to add **Heads** until all are complete.

That configures the SignalLCC-32H for a single head mast. Intermixed in these steps (including some parts that we skipped over)are some that are strictly electrical (Head Position, Lamp Intensity, Lamp fade) and some that relate to the signaling logic to be discussed later. While this part is not very complicated (with a road map like above) it is very tedious because of the repetition.

## Skipped Fields in Configuration

Less used fields that are in the Configuration were slipped in the previous steps Most important of these is the **Link Address** field. This is the address of the **Track Circuit** linked to the **Head**. This address will be needed in the logic that defines the Rule for the next Head (or previous, depending on how you count.)

Following the Link Address are several fields that define the way the lamps behave. Lamp Fade is an effect to mimic the fading of incandescent lights, Lamp

Intensity is the brightness of the lamps. These can be set after the signals are tested.

Further down the window, two events are created by the processing, **Aspect Set** and **Aspect Cleared** are reports that this **Aspect** has changed. These events are convenient for some purposes, but probably not common.

### Configuring a Dual Head Mast

Many railroad Rule Books include Rules for Masts with two Heads. These are used when a track splits off from the current track. The Rules for dual-head masts link the two heads into what operates as a single aspect. To support this in the SignalLCC-32H, the third option in the **Signal Processing** pull-down list is **Link to Previous**. This option combines the two Heads into one Aspect. To support two heads, the number of **Rules** is doubled (becomes 16).

To configure these two heads, first proceed as before to configure the top **Head**. This will be the **Head** that specifies the **Track Circuit**, the **Events** and other settings. It is the Mast definition. The next **Head** is the **Lower Head**.

Configure the **Lower Head** by selecting the next **Head** and specifying **Link to Previous**. Specify the Position to tell the board where to place the head data in the interface. (Note that the position is not necessarily the next position.)

Many of the next fields are not used in a dual head configuration. When you get down to the **Rules**, the top **Head's Rules** are used by all **Heads** Linked to that **Head**. Step through each **Rule** and assign the Lamps for this **Head's** appearance when displaying that **Rule**.

Thee next part of this chapter will detail the logic that can be used to support this many signal heads. LCC has the advantage that the logic does not have to be with the node that displays the signal aspect, so the next section is about the implementation of logic based in the TowerLCC+Q.

## TowerLCC+Q and Statement List (STL)

Signaling is complex. Every situation needs to be evaluated and designed for proper signaling. As noted before, prototype railroads devote entire departments to signaling. For them, it is a matter of safety. Train operators must learn all of the Rules before they can correctly operate on the railroad. The Operations Rule Book is the most important document for train operations. It defines the actions to take when encountering a specific signal aspect.

Personnel safety is not a real concern for Model Railroad signaling. Model Railroaders normally are of two minds about signaling—either they just want the impression that there are signals (aka "eye candy") or they are attempting to mimic a prototype. In some cases, signaling is actually an aid to running model trains, because it provides information when the operator is unable to see the situation. If "eye candy" is all you want, skip the rest of this chapter, because there will be some simple programming involved and this is getting complex.

To find enough memory to contain "Conditionals" a new board was designed. Using a newer micro-controller with the same architecture, a new TowerLCC was created. With increased memory, this micro-controller supports a low-level language called Statement List (STL). This language was developed

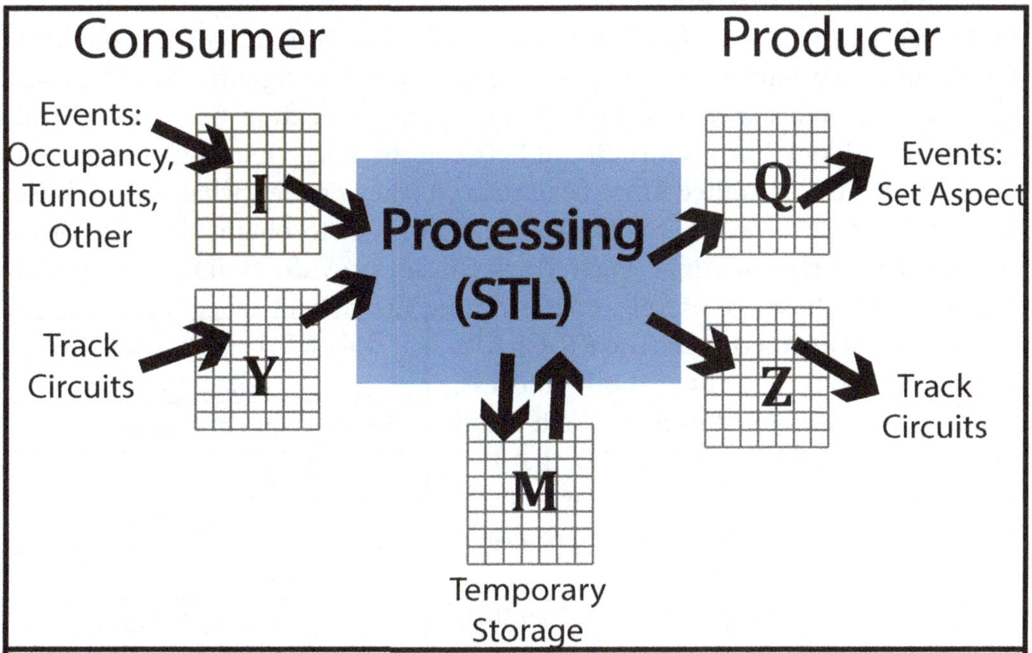


**Figure 8.9 STL Processing Takes in Events and Sends Out Signal Aspects as Events.**

for use in automated manufacturing and is oriented towards evaluation of logic expressions.

## Integrating STL with LCC

The STL language is designed to control Programable Logic Controllers (PLC). In PLC use, the inputs and outputs are usually directly connected to I/O lines. To integrate LCC's event-driven structure with STL, a supporting interface is needed. Primarily, STL provides the logic evaluation for LCC, so the LCC node needs to provide the inputs and outputs for STL. In the Producer/Consumer model of LCC, state changes create events which in turn cause state changes somewhere else. Receiving an event for STL causes a state change in the value of a bit in the specified memory location. This provides the needed inputs for signaling—Occupancy, Turnout Position and any other input. STL outputs are also stored in a table in memory. When a bit changes in the output tables, an event is generated on the LCC network to announce that change. In this action, STL becomes a Producer. These tables are shown in Figure 8.9. To reference the stored variables in the STL logic, the letter of the memory table is followed by the "row" and bit offset. An entry like Q2.0 means the third row, first bit in table Q. Similarly, there is a temporary memory which is used for intermediate values. It is referenced in the same way, like M0.5 as an example.

## A short Discussion about Track Circuits

Prototype railroads were faced with the problem of communicating signal information. They realized that they had a connection through the rails that could be used. The first use was to provide occupancy detection because the train wheels would look like a short circuit between the rails. Unoccupied track sections could carry a limited amount of data, so they created what we call Track Circuits. In LCC, Track Circuits provide a shortcut to obtain the state of the signal they are representing. Each Track Circuit provides 8 different aspect to choose from. These are useful to determining the correct aspect for the signal that is in the previous block. The RR-CirKits implementation requires a Link address, which is the basis for 8 events. Each event notifies the Track Circuit Receiver that this aspect is now displayed. It also clears the old aspect, so only one aspect can be present at any time.

Track Circuits are referenced slightly differently than input or output memory. To set up the matching events, the Track Circuits only need a Link that defines the base for the events. This base defines the row of the memory table to store the event state.

**Table 8.1 – Track Circuit Aspects**

| Offset | Aspect Name |
|---|---|
| 0 | Stop |
| 1 | Approach Medium |
| 2 | Approach |
| 3 | Approach Limited |
| 4 | Advance Approach |
| 5 | Approach Slow |
| 6 | Tumble Down |
| 7 | Clear |

An offset in the row designates the "Speed" as shown in Table 8.1. The **EventID** is the sum of the base and the offset. For example, setting the Approach aspect with a Link base of "**02.01.57.11.00.7C.01.68**" would send an eventID "**02.01.57.11.00.7C.01.6A**".

The Link is entered (probably Pasted) into one of the Track Receiver or Transmitter Circuits (0 - 15) to link the track circuit to a specific set of events. Then, if this is circuit 2 , setting an Approach aspect would mean storing a True (1 bit) in Z2.2.

You may have noticed that there are multiple places to Link a Track Circuit in the boards discussed in this chapter. Specifically, both the TowerLCC+Q and the SignalLCC-32H have Track Circuit Link fields. To avoid confusion, I suggest that only one of these be used, probably the SignalLCC Links. These will be set regardless of the setting in the TowerLCC+Q, and are likely to be more consistent with the aspect that is set. However, if you are setting the mast with some other method, you can always use these Track Circuits.

## I/O Summarized

STL needs LCC inputs, so events are converted to Boolean states in Input Tables. Outputs from the STL logic are stored in an Output Table. When the processing detects a change in an output Boolean variable, an event is created to signal that change. STL logic references these variables in the form Ar.o, where A is the table, r is the Row or byte, and o is the offset in the byte. Using these methods, STL is provided an interface to the LCC model of Consumers and Producers.

## STL Processing Overview

To understand the Statements and actions in the next few sections, a high-level overview of the processing is needed. The previous sections on I/O left us with a set of tables in memory, each with a specific purpose—Input(I), Output(Q), Track Receiver(Y) and Track Transmitter(Z). A temporary memory(M) was added to store intermediate values. Events from the LCC network set or clear bits in the Input memory table; Every 20 milliseconds, STL processing sets or clears bits in the Output memory table based on the logic statements. After processing the logic, the Output table is checked to find changes. If a change has oc-

**Expressions, Statements, or Networks?**

*In the TowerLCC manual and Seimans documentation the definitions are somewhat fluid. To provide some consistency, this book will refer to an Expression as the item which is compiled as a unit. Another name that is seen is "MultiLine." this is just the space provided to hold Expressions. The compiler does not stop at the end of a multiline; It just moves to the next available multiline.*

*Certain operators terminate expressions and assumptions drive the state of the expressions at the beginning of each one. I will attempt to make this clear as we do.*

curred, an event is sent to the LCC network to announce that change. Details of the processing are explained below.

## The STL Processing Language

The STL language used in the TowerLCC+Q is a small subset of the official STL language. It is limited to Bit operators and uses the variable designations described above. With these limitations, the subset of STL is very efficient language that executes quickly.

STL is one of a class of languages that are called Assembly languages. Assembly languages mimic the architecture of the processor that is used for execution. STL is designed for a type of processor that requires a variable loaded into a specific location–often called an accumulator–before it can be used in an operation like addition, Boolean AND, etc. In hardware this limits the data paths to and from memory. In STL it means that all operator statements only reference one memory location which normally combine with the result of the previous operation, called the RLO (Result of Logic Operations). Because of this, each instruction in STL starts with an Operator, like A(and) or O(or). To reduce the memory required to hold the operations, a shorthand version of each is used, so And becomes just A, Or becomes just O and so on.

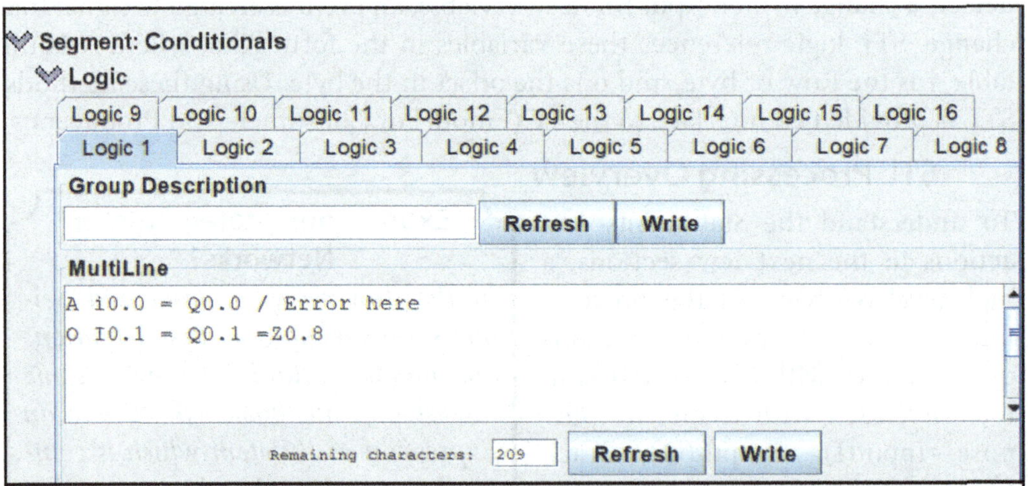


**Figure 8.10 STL Expressions are entered into MultiLine Blocks in the Conditionals Segment**

In the LCC implementation of STL, the language expressions are entered into space in the Conditionals Segment named "MultiLine." 16 of these blocks (called "Logics" in the Conditionals segment) are available,; each can hold up to 255 characters. As seen in Figure 8.10, each block can be named to identify the purpose of these expressions. In the TowerLCC+Q, The Compiler (or more precisely, Assembler) compiles (Assembles) the executable program every time the board is initialized. This means that the board must be rebooted to get a compiled version of the expressions, and it is that step that checks the expressions for errors. Like any good compiler the STL compiler ignores white space (blanks and new lines) and allows comments. Comments are entered either between the "/*" and "*/" characters within a line, or by indicating the rest of a line is a comment by inserting a double slash "//". As an example, a line might look like this:

**A I0.0 /* loads the RLO */ A I0.1 = Q0.0 // And two inputs for output**

Syntax errors (errors in writing the instructions) are presented in the Syntax Messages Segment, if you look for them. When you open the Syntax Messages Segment the message fields display the last message, not the current. To see the current message,1 you must Refresh each field. The segment would then look like Figure 8.11 on page 149. In this case there are two syntax errors, one

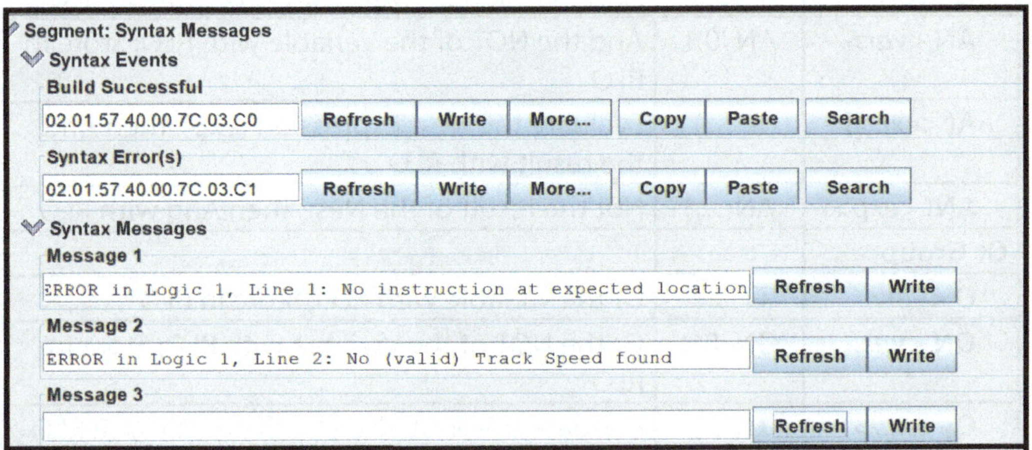


**Figure 8.11 Syntax Errors are Displayed in tis Segment**

in each line. The first is because there is only one "/" before the comment on line 1; the second is because the **Track Transmitter** reference is incorrect, it is out of the 0-7 range.

Notice the presence of the two **EventIDs** available to signal the results of the compiler. The first is sent when the build is successful, i.e. no errors were found. The second is sent when errors are reported. Using these Events provides an instantaneous report from the compiler.

## STL Operators—Boolean Bit Logic

The basic Boolean operators are implemented with two additional variants, an inversion on the input and a nesting operation. To review the basic Boolean Logic operators see Figure 7.1 on page 114. The three basic operators can be augmented with either an N for NOT or a parenthesis for nesting.

If the operator is modified with an N, that indicates that the variable designated will be inverted (1 becomes 0, 0 becomes 1) before the logic is evaluated.

If the operator is followed by a parentheses, a Nesting occurs. This means that the current state of the expression is stored on a stack and a new expression is started. As before, the RLO is loaded with the variable designated and evaluation continues until a closing parentheses is encountered. When that is found, the state stored on the stack is retrieved and the RLO from the "nest" is used as one variable for the "Nesting" operator in combination with the saved RLO. That sets the RLO for the rest of the expression. Nesting is limited to 7 levels.

The Boolean Operators are defined in Table 8.2.

**Table 8.3 – Boolean Operators**

| Operator | Example | Action |
|---|---|---|
| **AND Group** | | |
| A <*var*> | A I0.0 | And the Variable (var) with RLO; Store in RLO |
| AN <*var*> | AN I0.1 | And the NOT of the Variable with RLO; Store in RLO |
| A( <*exp*>) | A(…) | Evaluate the Nest Expression (exp), then And the result with RLO |
| AN( <*exp*>) | AN(…) | Not the result of the Nest, then And with RLO |
| **Or Group** | | |
| O <*var*> | O I0.2 | Or the Variable with RLO; Store in RLO |
| ON <*var*> | ON I0.3 | Or the NOT of the Variable with RLO; Store in RLO |
| O( <*exp*>) | O(…) | Evaluate the Nest, then Or the result with RLO |
| ON( <*exp*>) | ON(…) | Not the result of the Nest, then Or with RLO |
| **XOR Group** | | |
| X <*var*> | X I0.4 | Xor the Variable with RLO; Store in RLO |
| XN <*var*> | XN I0.5 | Xor the NOT of the Variable with RLO; Store in RLO |
| X( <*exp*>) | X(…) | Evaluate the Nest, then Xor the result with RLO |
| XN( <*exp*>) | XN(…) | Not the result of the Nest, then Xor with RLO |

**Table 8.2 – Expression Termination**

| Operator | Example | Action |
|---|---|---|
| = <var> (Assign) | = Q0.1 | Stores RLO in the designated bit; Multiple Assigns are allowed in an expression; |
| R <var> | R Q0.2 | Resets the designated bit if RLO = 1 |
| S <var> | S Q0.3 | Sets the designated bit if RLO = 1 |

## STL Operators—Expression Termination

Each expression terminates with an assignment to an output table location. The Assign operator is simply an Equal sign (=) with the location designation following it. The Assign operator stores the RLO at the designated location. A very simple example would be:

**O I0.0 A( x I0.1 A I0.2) = Q1.0 // Test RLO**

In this example, the output in Q1.0 only sends an event if all the inputs are set to 1. Tracing the action:

1. **O I0.0** sets **RLO** to 1;
2. **A(** starts the Nest by storing the state;
3. **X I0.1** puts a 1 in the Nest's **RLO**;
4. **A I0.2** ands the values in **I0.1** and **I0.2** to put a 1 in the Nest **RLO**;
5. ) Closes the Nest and performs an And with the original **RLO**;
6. = **Q1.0** then stores the **RLO** in **Q1.0**, and sends an event if that is a change.

## STL Operators—RLO Operators

Several operators use or change the RLO directly. These operators are defined in Table 8.4. These are unary operators, which means they do not have a variable specification. The action is defined to use the RLO bit. In the special case of SAVE, The RLO is saved in the BR bit where it can be used to start the next expression.

**Table 8.4 – RLO Operators**

| Operator | Example | Action |
|---|---|---|
| NOT | NOT | Inverts the RLO, 0 becomes 1, 1 becomes 0 |
| SET | SET | Sets the RLO; RLO becomes 1 |
| CLR | CLR | Clears the RLO; RLO becomes 0 |
| SAVE | SAVE | Saves the RLO for use in the next expression |

As an example of using the SAVE operator, the expression below performs a SAVE at the end of the first expression and then recovers that saved RLO at the beginning of the next expression.

**A I0.0 A I0.1 SAVE =Q1.0**

**O BR A i0.2 =Q1.1**

## STL Operators—Edge Transitions

It is sometimes necessary to know when a input has changed. For example, a change in a Block Occupancy Detector indicates that a train has either entered or left the block. For this reason, these operators are available. Any input can be tested by these operators. As shown in Table 8.5, the operators need a memory variable in the M table to use to remember the previous RLO. These edge operators detect the edge on the preceding RLO. The memory designation is selected by the user and must be reserved for this purpose. Each processing cycle that executes these operators will save the current RLO in memory. The next cycle that executes the operation will check to determine if a change has occurred and output a 1 of it has changed and a 0 if it stayed the same. Because of this, the 1 output is most likely only one processing cycle long.

**Table 8.5 – Edge Transition Operators**

| Operator | Example | Action |
|---|---|---|
| **FN** <*var*> | A I0.0 FN M1.0 | Detect Negative edge; Input transitioned from 1 to 0 since the last processing cycle |
| **FP** <*var*> | A I0.1 FP M1.1 | Detect Positive edge; Input transitioned from 0 to 1 since the last processing cycle. |

## STL Operators—Logic Control

Jump instructions allow you to alter the flow of logic instructions. This is useful to provide different outputs based on the same starting logic. Jumps in the TowerLCC+Q STL can only go forward. To identify the destination of the Jump, a

**Table 8.6 – Jump Operators**

| Operator | Example | Action |
|---|---|---|
| *label:* | Exit: | Jump Labels are 1 to 4 characters followed by a colon; The first character must be a letter |
| **JU** <*label*> | JU EXIT | Jump unconditional to label; Only forward jumps are allowed |
| **JC** <*label*> | JC EXIT | Jump if RLO=1; If RLO=0, no jump occurs and RLO is set to 1 |
| **JCN** <*label*> | JCN EXIT | Jump if RLO=0; If RLO=1, no jump occurs |
| **JCB** <*label*> | JCB EXIT | Jump if RLO=1; If RLO=0, no jump occurs and RLO is set to 1; In either case, copy RLO into BR |
| **JNB** <*label*> | JNB EXIT | Jump if RLO=0; If RLO=1, no jump occurs; In either case, copy RLO into BR |
| **JBI** <*label*> | JBI EXIT | Jump if BR=1; If BR-0,no jump occurs and RLO is set to 1 |
| **JBNI** <*label*> | JBNI EXIT | Jump if BR=0; Ifd BR=1, no jump occursand RLO is set to 1, |

label is entered at that point. Labels are 1 to 4 characters where the first must be a letter. The label is followed by a colon to identify it as a label. Jump operators are either unconditional or dependent on the RLO or BR. The Jump operators are shown in Table 8.6.

Several Jump instructions modify either RLO or BR. This is so the following expressions can operate with known RLO values. The expressions following the jump can start without loading a new value.

## STL Operators—Timer Operations

The TowerLCC+Q STL includes 64 Timers. These timers are referenced as variables T0 to T63. These timers can be used as variables in STL expressions. The value returned is a 1 if the timer has not expired and a 0 if it has reached the time interval specified.

Timer operations are triggered by RLO edge transitions, usually from 0 to 1. When the transition is detected, the timer operation is activated. The result of this is specified by the type of timer operation as defined below.

| Table 8.7 – Timer Base Values | | |
|---|---|---|
| **Base** | **Interval** | **Range of Time (1 to 999)** |
| 0 | 10 msec | 10 msec to 9 s, 990 msec |
| 1 | 100 msec | 100 msec to 1mpn, 39.1msec |
| 2 | 1 sec | 1sec to16min, 39 sec |
| 3 | 10 sec | 10 sec to 2 hr, 46.5 min |

To initialize the timer, a Load instruction must be entered. The Load instruction has the form "L W #t #xyz" where t is the time base is 0 to 3 as shown in Table 8.7. The "xyz" value is the multiplier to define the total timer interval. Leading blanks are ignored.

Since the processing cycle is 20 msec, timer intervals of less than 20 msec will be undetectable as less than 20 msec. If an interval less than 20 ms is specified, the timer will not be checked until the next processing cycle 20 ms later. Therefore, the real lowest resolution of the timer is 20 ms, due to the processing cycle time. In use, an event to start the timer can come at any time in the 20 msec. Cycle time, then the timer will start at the next processing cycle, sometime after the beginning of the cycle. If the timer is set at 20 msec, the timer will not complete until sometime into the next cycle, which means the event for completion is not processed until some time into what is now the third 20 msec cycle. When you add the time for propagation of the events, it is impossible to predict a resolution between events less than approximately 100 msec. Therefore do not use this for precise measurement of time.

Timer Operators require that the timer be initialized before any operator can start the timer. Timer operators encountered before the initializing Load will be ignored until the next processing cycle. To select the timer in the range from 0 to 63, the operation immediately following the Load must be a Start operator specifying the timer that you want to initialize. Think of the Load and Start operations as set up for the Timer for the time period. In my tests, release V1.10 of the TowerLCC+Q does not support subsequent Start operations on the same timer without a Load. There is no implemented way to re-trigger the timer to lengthen the time. If you take the trigger (RLO) to 0 and back, it stops the timer and restarts it with a full time periods. Timer operations are somewhat tricky. The only way I have found to reliably run a timer is to use only one timer, in only one location in the logic. The timer can be restarted, multiple timers are okay, but only one instance of each.

## A Demonstration Layout as an Example

To tie all of thus together, I present the following example. I will identify the steps and show the Configuration Details for this example, although I have not actually built this, because it is too large. It includes many of the features necessary for one section of the Belmont shore railroad, so i am planning to implement much of this in the layout. What I am doing in this section is essentially a prototype of the logic that will be needed.

The basic track plan is a dual track mainline coming into a medium sized town with a passing siding running between the mainline tracks through the town. On both mainlines there are several stub tracks to service industries. The industries include an icing siding for reefers, a brewery, an orange grove and a stockyard. At the ends of the town the dual mainline has the ability to enter the passing siding. This becomes similar to the dual track mainline entering a single track two-way section.

Additional concerns for signaling are the industry sidings, which may produce essentially unit trains filled with oranges or cattle. These need to be treated as little branch lines entering the mainline. That adds signals to the example.

This example includes the two ends with wye turnouts to enter the passing siding, two industry service tracks, one for the icing platforms and one for the orange grove. The icing platforms are on a full siding with entry/exit on both ends; the orange grove is a long stub out to the orange groves. A conceptual Control Panel is shown in Figure 8.12.

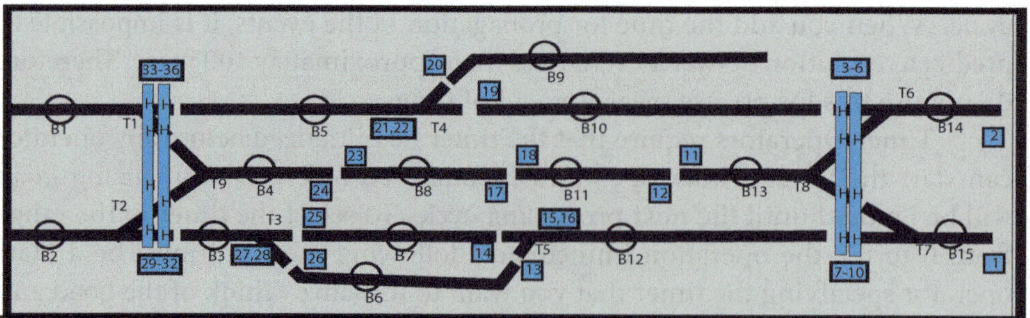


**Figure 8.12 The Control Panel identifies the Inputs and Outputs for Configuration.**

To prepare for the configuration and logic development, the first strep is to identify the inputs and outputs. The Block Detection inputs are shown on Figure 8.12 as circles around the track blocks. Block occupancy is labeled from B1 to B15. Turnouts are both inputs and outputs. Since the Kato turnabouts this is designed around do not have switches to sense the position, the pushbutton inputs that control the turnouts will be used to indicate a change in position. These pushbuttons will share lines with the turnout drivers to minimize nodes. As an initial design, plan on hooking up all block detectors to a TowerLCC+Q and the turnouts to a SignalLCC-32H. The unused line on the TowerLCC will be used to control the extra turnout from the SignalLCC. Using the same plan for the STL inputs, we will configure B1 → Line1 → I0,1 using Line1's events. Unfortunately, more than one SignalLCC will be needed.

**About Kato Turnouts**

*Kato Turnouts use a single solenoid to switch the points. This means that the voltage to change the points must be reversed. This is similar to changing direction for a DC motor. The common way to control motors is with something called an H-bridge. Leaving those details to a later Appendix, it also should be noted that the voltage should only be applied for a short time, like 20 msec. All of this results in a special driver board which will be described in Appendix B.*

The first issue to be noticed on this layout design is the three turnouts at each end that lead to the center passing track. Two things need to be considered because of this design. First, the wye turnout control is coupled to the control of the turnouts leading to it. That means that the wye does not require an independent button or switch, because it can be changed when the leading turnout is changed. Second, there is a situation where both leading turnouts could compete for the wye. If that were to happen, a derailment and/or short could occur. Some form of lockout protection would be desired.

This is a great opportunity to try out our new STL logic. If we want to prevent the issue, we can lockout one leading turnout if the other one already has changed to use the wye. To show the options, the "Truth Table", Table 8.8, defines the needed logic. In this table, the first two columns represent the toggle switches used to select the track that transitions to

**Table 8.8 – Wye Turnout Truth Table**

| Toggles | | Turnouts | | | LEDs | | Event |
|---|---|---|---|---|---|---|---|
| S1 | S2 | T1 | T2 | T9 | L1 | L2 | |
| 0 | 0 | N | N | N/C | Off | Off | D0 |
| 0 | 1 | D | N | L | Off | On | D2 |
| 1 | 0 | N | D | R | On | Off | D4 |
| 1 | 0>1 | N/C | N/C | N/C | N/C | Flsh | D2/D3 |
| 0>1 | 1 | N/C | N/C | N/C | Flsh | N/C | D4/D5 |
| 1 | 1 | N/C | N/C | N/C | N/C | N/C | – |

the Passing siding, the next three columns are the outputs to the turnout drivers, and the last two columns are outputs to the LEDs that show the direction beside the first two turnouts. Turnout direction is Normal (N), Diverging (D) or N0 Change (N/C) and the LED state is On, Off or Flsh (Flashing).

Examination of the table reveals dependencies where the output follows another output. An example is the LED outputs which are the same as the T1 and T2 outputs. Because of this, one event can be used for both outputs. Using the same event for the three easy cases will save space in the Logic. In those cas-

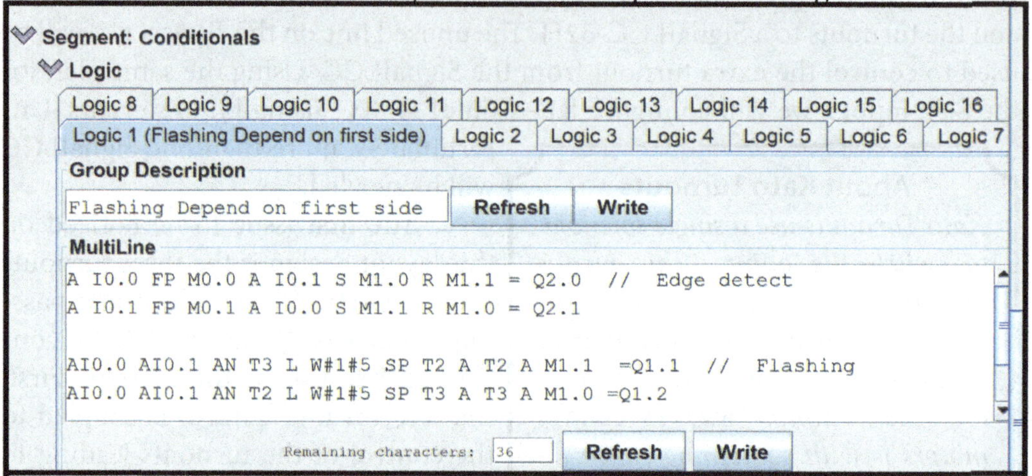


**Figure 8.13 Logic Expressions to make a Flashing LED**

es where the action is the opposite, the event will be used to turn off (or on) the output. For reference, the last column identifies the last two characters in the EventIDs for each logic expression.

The tricky part of these outputs is the flashing outputs at the bottom. These use two of the operators that are not simple logic operators, Edge detection and Timer operations. The Logic expressions are shown in Figure 8.13. The first two lines detect which Toggle switch was the second one. This is achieved by the detection of a rising edge on the I0.0 input after I0.1 is on. Simultaneous actions would have to be within the 20 msec processing cycle and would then depend on which one was processed first. The logic expression **A I0.0 FP M0.0 A I0.1** is True if input I0.1 is true and I0.0 is a rising edge. When that is true, a memory bit, M1.1 is set, the bit for the other toggle is reset and the output is sent. Memory bits are Set and Reset because the next processing cycle will evaluate this expression as false, so the Output to Q3.0 will become false.

The two "Flashing" expressions form a loop which is controlled by the Input variables. Each expression starts a timer using the rising edge of the "trigger" expression. The trigger expression **AI0.0 AI0.1 AN T3** Starts the Timer T2 when it becomes True. I0.0 and I0.1 are both the enabling inputs and the stoppers to stop the loop. Timers T2 and T3 set the duration of the blinking, and the LED selection is determined by the expression following the Timer.

This method may be useful for LEDs that are sometimes flashing and sometimes constant. It works because the "Processing Cycle" is itself an infinites loop. Looping over the expressions and allowing tests of"global" variables lets you set a variable in one place and use it in the next cycle.

The second example from this layout is the center passing siding. As shown, this siding can be entered from either end and is essentially a single track, two-way siding. Because of its length, it has to be treated as a single track section between two dual track mainlines. Prototype railroads used APB signaling in cases like this to protect two-way movements. The technique is called

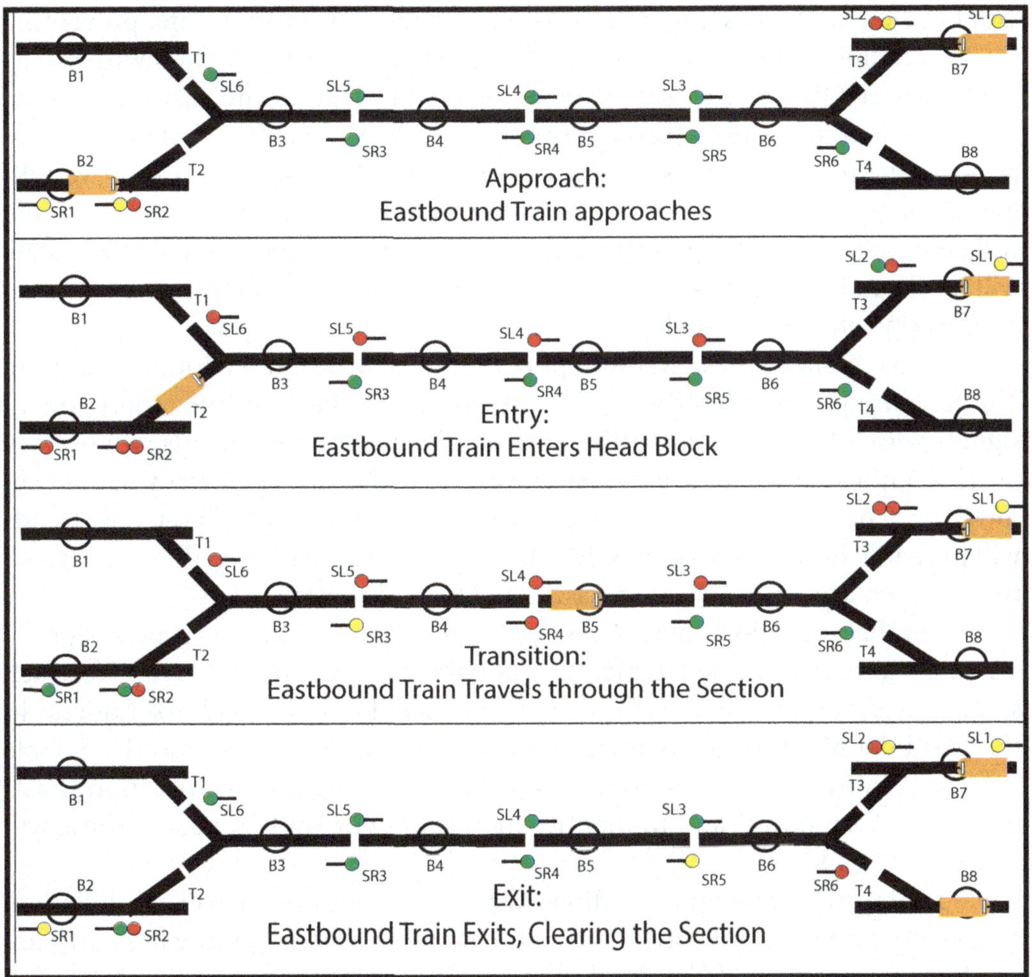


**Figure 8.14 States of the Tumble Down Signaling**

"tumble-down" signals, named for the semaphore signals that first used it. At one end of the single track, a train would enter, causing all of the signals to trains moving in the to drop down onto the Stop aspect, therefore, they would "tumble down." Figure 8.14 shows the progress of a train as it travels through the tumble down section. In this situation, two trains have approached the single track section. The Eastbound train will acquire the section by passing into the block labeled B3 first. During this Approach, both ends have aligned their turnouts to enter the middle passing track. The signals on both end, SR2 and SL2, show an Approach Aspect, alerting the engineer to approach the single track prepared to stop.

The next state, Entry, shows the Eastbound train occupying both block B2 and B3. By using the Edge Detection on block B3, the direction of travel can be determined. Using the Direction, a "Tumble Down" Aspect is set in the Track Circuit for signal SL5. Track Circuits for all signals back to SL3 sequentially set Tumble Down. SL2 is set to a Stop Aspect for the Diverging Route, preventing the Westbound train from entering. All the signals from SL2 to SL5 will display a Stop aspect. On the other side, everything is still Clear for the Eastbound train.

While transitioning through the section, the signals for Eastbound traffic perform the normal actions for ABS to prevent following trains from overtaking. As long as there is a train occupying blocks B3 to B6, the Tumble;e Down will continue to show Stop. As the Eastbound train exits, the Tumble Down state will be cleared allowing the Westbound train to enter and the process is started on the opposite set of signals.

Implementing these states requires some planning and a lot of repetition. The first thing is to identify the EventIDs for the signal heads. There are 12 signals with 14 heads on just this short section Many of the heads will be very nearly identical in the logic expression with the differences primarily in the EvemtIDs. Each head aspect will be set by an Event sent from the STL Logic. Heads will probably have 3 to 5 aspects. Identifying all of the aspects will be very repetitious. Each aspect will need an associated Q Table entry.

On the input side, there are 8 occupancy detectors and 4 turnout signals. These are allocated in the I Table. Additionally, there are track Circuits for each signal. These are in the Y table; 12 of the 16 will be used; the Link Address in the Head definition of the SignalLCC-32H needs to be copied into the Y table entries. Finally, there will be some temporary bits to define the direction (Eastbound or Westbound) and to identity the Leading or trailing edges. These will be in the M table.

Setting all of this up is a tedious task and it is unique to every installation. I cannot provide your EventIDs, so in the following discussion the logic will use symbolic names taken from the drawing in Figure 8.14. I recommend that you create a cross reference table to define the STL table entries.

As before, the signals are defined from the end to the beginning. This is because the Track Circuits are filled in that way to provide the "Following" protection. To start looking at the logic, signal SR4 will be used as an example. SR4 has two primary input sources, occupancy of block B5 and the Track Circuit for SR5. There are four output aspects: Tumble Down, Stop, Approach and Clear. In words: If the next signal is Tumble Down, set Tumble Down; If Block is occupied, set Stop; If next signal is Stop, set Approach; and If none of the above, set Clear. In STL this is:

**A Y{SR5}.6 R Q{SA} R Q{AA} R{CA} =Q{TA}**// Test for Tumble Down
**JCB EXIT**
**A{B5} R Q{AA} R Q{CA} = Q{SA}** // Test for Occupancy => STOP
**JCB EXIT**
**A Y{SR5}.0 =Q{AA}** // Test for next signal Stop => Approach
**JCB EXIT**
**=Q{CA}** // Set for Clear aspect since no other has set
**EXIT: =Q1.0**

To use this fragment of logic, the items enclosed in braces (aka curly brackets: {}) need to be pasted in from your configuration. The next Track Circuit is in the Y table at SR5. The outputs are the Aspects TA (Tumble Down), SA (Stop), AA (Approach) and CA (Clear).

The pattern of processing is to test for the condition (like occupancy) and then set the Aspect event. As the aspects are set, the output variable in Q is reset so that the next time it is set, it will generate the event.

To understand this, the first line is the test and set for the Tumble Down state. The test is on the sixth bit of the Track Circuit from signal SR5 – **A Y{SR%}.6**. If this is true, the output to set the Stop aspect is reset – **R Q{SA}**, the Output for Approach is reset – **R Q{AA}**, the output for the Clear aspect is reset – **R Q{CA}**. The last operation in this line is to assign the RLO, either True or False, to the Tumble Down aspect – **=Q{TA}** The next line jumps to the **EXIT** of this fragment if RLO is True. If not, the next test is tried, having left the **Q{TA}** with a 0 in it.

Implementing the logic for the first block (aka the Head Block) in the Tumble Down section requires determining the direction of travel. In our example, the Head Block is block B3. The Eastbound train enters that bloc to set all the Westbound signals (SL2 through SL5) to the Tumble Down Aspect. The propagation of the Tumble Down is performed by the logic using the Track Circuit shown above. The Head Block signal, SL5 will be modified by replacing the first line with logic that sets the Tumble Down and will reset it when the train exits the section. That logic depends on a memory variable to know which set of signals need to be Tumbled Down. The logic for signal S:5 is:

**A M{Eastward} A( O{B3} O{B4} O{B5} O{B6}) R Q{AA} R{CA} =Q{TA}**

This logic tests to see the direction of travel and tests occupancy of the section. It will automatically reset the Tumble Down when all trains have cleared the section. This allows following trains to enter while the first train is still traversing the section. The only thing left is to set the direction ion **M{Eastward}.** The logic for that is not associated with a signal directly, so it only sets a memory variable. It should look like this for both setting and clearing the Tumbe Down:

**A{B2} A{B3} FP M0.0 S M{Eastward } // Set Direction**
**A M{Eastward} AN I{B3} AN I{B4} AN I{B5} AN I{B6}R M{Eastward}**

**// Release Direction**

The first expression sets the direction by testing the rising input from block B3 while the train is entering from block B2. The result is used to "Set" the bit until the next expression resets it. Set is used because the next pass through this expression will be false because the input is no longer rising.

The second expression reset the direction bit by testing if there is occupancy in the section. The set of And Not combine with the direction bi to reset the direction bit. Since the direction bit causes the Head Block signal to set the Tumble Down aspect, this bit turns the Tumble Down on and off.

> ***Example Testing***
>
> *The Examples in this section have been simulated in a TowerLCC+Q. Each one has been compiled several times to make certain the syntax is correct. Unfortunately, because events are only created when the output changes, it is difficult to see what is happening inside the logic. This process revealed several mistakes in the current (Sept 2024) manual. There are also some things that have not yet been implemented. Watch for updates of both the firmware and the manual.*

This logic is repeated for entry from blocks B1,B7 and B8, with the appropriate direction bits. You may have to add more intermediate blocks and make some other changes. This is just a starting point.

## STL Summery

STL programs are collections of Expressions that exist in the LCC environment. As noted above, each operation starts with a Operator like A or O, followed by a variable. Definition of the Operators was covered in previous sections. The important point about the Operators is that they only identify one variable at a time, using the result of the previous operation as one of the variables in an operation. This is called the Result of Logic Operation (RLO) and it is kept internally in a Status word. As the name implies, the value stored in the RLO is the result from the previous operation. The tricky part is where does this come from in the very first logic operation? At the beginning of the processing the RLO is initialized by simply loading RLO with the variable specified by the first operation. As an example, the operation **A I0.0** means to AND the RLO with the value found in I0.0 may be the first operation. If so, the A is ignored and the value in I0.0 is loaded into RLO. It doesn't matter what the Operand is, the variable is loaded into RLO. Given this behavior, I recommend you start each

> **Applicability**
>
> *This document is based on TowerLCC+Q firmware release 1-10 as described in the manual release D. Future releases of either may change some of the details. Always check the release notes to determine what may have changed.*

expression with an AND operand,However, there is a way to pass the RLO from one Expression to the next. Using the SAVE operator.

## Important points to remember

- STL is a Logic language for bit operators.
- Inputs are set into memory tables  by events received.
- Outputs that have changed in the output tables create events.
- Logic is entered into Multiline block's in Conditionals
- To compile the Logic, the node containing it must be rebooted.
- Syntax Errors are displayed after a Refresh in the Syntax Messages Segment
- Operators combine an Input variable from tables I, Y or M with the RLO (Result of Logic Operation)
- The first Operator (e.g. A, O, X, An) in a statement is ignored—the referenced variable is loaded into RLO.
- A STL statement is terminated by an Assign (=, S, R, or SAVE).
- The RLO is cleared at the end of each expression unless it is SAVE.
- Expressions are processed every 20 milliseconds, then the Outputs are checked for changes. If an output has changed, the associated event is generated.

STL is like a factory where events come to it, are processed into new events that are sent to the signal heads for display. The input tables are the receiving dock, the memory tables are internal shelves to store intermediate results and the output tables are the shipping area. The STL logic is the machinery that processes the events into new events. In fact, STL was created to emulate relay logic.

# Chapter 9 —LCC Fast Clocks

Time is an important part of any railroad. In addition to timetable operations, movement of trains is always concerned with how long the movement will take. In model railroads, the issue of scaling becomes a concern when attempting to make realistic moves. Distances between points of interest are short because of the scale and because model railroads typically use selective compression to eliminate the distances between those points of interest. The standard solution to making the time between points seem realistic is to operate with a fast clock.

A fast clock is sim[ply a clock whose minutes and hours are shorter than real minutes and hours. Fast Clocks determine the length of minutes by a ratio or rate. Rates are determined by the scale and distances on the specific layout. Wide variations exist, so each clock must by adaptable. Unlike real time, Fast Clocks need to be set to specific time values and may need to be paused during operation.

Incorporating a fast clock with the LCC network enables several capabilities. Simply providing a time to multiple points on the layout enables local displays for operators. Additional features can be used to time sequence layout structure and sequence animation. These features add significant realism to the layout.

This chapter begins with a summary of the LCC Standard that defines the Fast Clock interactions, the Broadcast Time Protocol. Then the LCC Fast Clock available from Logic Rail Technologies™ will be described. Displays of the Fast Clock time will also be described. The next chapter will describe the use of the time events that are generated for lighting effects.

# The Broadcast Time Protocol

Defining a set of common events for a Fast Clock provides the interface to LCC. To make a common, global definition of these events, four of the Well Known Events are used. These four EventIDs define the first 48 bits of the Broadcast Time Protocol events and provide for defined clock Types. The last 16 bits of these events is used to identify actions and transfer data to or from the Fast Clock. Because these EventIDs are unique, multiple fast clocks can exist on the same network. The four Clocks defined in the Standard are:

- 01.01.00.00.01.00 – Default Fast Clock
- 01.01.00.00.01.01 – Default Real-Time Clock
- 01.01.00.00.01.02 – Alternate Clock 1
- 01.01.00.00.01.03 – Alternate Clock 2

As you can see, the first 5 bytes (40 bits) are identical and the last byte defines the specific clock type. Although the clocks are named, there is no expectation that a clock is used for any specific purpose.

Two general types of events have been defined: Reports and Requests. Reports provide data to the network from the fast clock and requests command the fast clock to do some action.

## Events that Report Time

After displaying time at a specified rate, a fast clock's primary purpose is to tell everyone what time it is. In the Broadcast Time Protocol there are five events generated by the fast clocks to tell the LCC network what time it is. Table 9.1 shows these events and the variable bytes which carry the data. Note that in this table,all the data in bytes 7 and 8 are in hexadecimal. That is the way the events will look when seen in the LCC Traffic Monitor and how they must be encoded for the LCC network. More on this topic later. Details for each report event are:

**Table 9.1 – Broadcast Time Report Events**

| Event Name | Byte 7 | Byte 8 |
|---|---|---|
| Report Time Event | 0x00+hour<br>[0x00 - 0x17] | Minutes<br>[0x00 - 0x3B] |
| Report Date Event | 0x20 + Month<br>[0x21 - 0x2C] | Day<br>[0x01 - 0x1f] |
| Report Year Event | 0x3000 + Year<br>[03000 - 0x3FFF] | |
| Report Rate Event | 0x4000 + Rate*4<br>[0x4800 - 0x77FF] | |

***Report Time Event*** — The Report Time Event is the most used event in the protocol. Once the Fast Clock is started, The time changes frequently and these time events are used to sequence lights or synchronize other clocks. To perform these tasks, the Report Time Event is required to be sent at least once per real hour, but no more frequently than once per real minute. Specific time events are identified by the Consumer Event Identified message required by the Event Transport Protocol (see Chapter 1).

The Report Time Event is the simplest event because it is two bytes, one for the hour and one for the minute. Remembering that these are in hexadecimal, the ranges in Table 9.1 on page 164 equate to 24 hours and 60 minutes.

***Report Date Event*** — Date events are less frequent than time events and can often be ignored. Most model railroads do not have different actions for different days, at least not in the time periods that we operate. Regardless, the Broadcast Time Protocol includes a date report. Like the time event, bytes 7 and 8 are used for the two numbers that define a date (month and day). Note that the date does not include day-of-the-week. If combined with the year (in the next report) algorithms are available to calculate the weekday. This might be useful if you are modeling a church or school that only rings a bell on certain days, but few people will stand around long enough to observe that refinement.

***Report Year Event*** — Even less likely is the year event. For completeness, the Year is included. The range is set by the number of available bits. Starting at Year 0, the range extends to the year 4095.

***Report Rate Event*** — Everyone seems to have an opinion on the correct rate to run a Fast Clock. The Broadcast Time Protocol allows a large variation in rates to satisfy everyone. Rates go from very fast (8 hours per real minute) to very slow (1 "fast" clock minute every 4 real minute); Forward or backward (yes, you can go back in time.). Zero works to stop the clock, but it is easier to use the request in the next set.

The number in the event is four (4) times the rate of the Fast Clock. This multiplier allows a finer resolution of times, like 2.5 times, 8.25 times and so on. It also allows a "Slow Clock" running at ¼, ½ or ¾ speed.

Negative rates are represented in 2's complement form, which is the easier to use in programming. If you are not familiar with this form, the simplest way to get a number for use in this events to get the number from the Programming Calculator on your computer. Use the last three digits to get the correct negative number. Then ask yourself "What am I really doing that needs a clock running backwards?"

The events that request changes to the clock are similar to the events that report time. The additional events in Table 9.2 are used to control the operation of the clock. These requests can be sent when the clock is stopped or running.

### Events to Control the Clock

| Table 9.2 – Broadcast Time Request Events | | |
|---|---|---|
| **Event Name** | **Byte 7** | **Byte 8** |
| Set Time Event | 0x80+hour<br>[0x80 - 0x97] | Minutes<br>[0x00 - 0x3B] |
| Set Date Event | 0xA0 + Month<br>[0xA1 - 0xAC] | Day<br>[0x01 - 0x1f] |
| Set Year Event | 0xB000 + Year<br>[0xB000 - 0xBFFF] | |
| Set Rate Event | 0xC000 + Rate*4<br>[0xC800 - 0xC7FF | |
| Query Event | 0xF000<br>Reports Current State | |
| Start Event | 0xF001 | |
| Stop Event | 0xF002 | |

***Query Event*** — When the clock receives this event, it responds by sending each one of the Report events to report its current state. The clock continues to run, so this is just a snapshot at the instant of the request.

***Start Event*** — This event is used in combination with the ***Stop Event*** to control the clock's run state. When the clock receives this event it starts running at the current time and rate.

***Stop Event*** — Requesting the clock to pause is an important feature. In Model Railroading, breakdowns and other happenings require a pause in the action. Using ***Stop and Start*** allows these breaks in the action.

# Fast Clock Implementation

Once the protocol is established for LCC, many options start to fall into place. This section considers the options for implementation. The next section will describe the Logic Rail Fast Clock. We will see how the choices match with the options described here.

In the protocol there are four defined clocks that have specific EventIDs. There is no requirement to implement any specific clock, nor is there a requirement to implement more than one. The option may be chosen to implement just the Default Fast Clock and no other. Alternately, all four clocks could be implemented, all running simultaneously. Implementing just one and allowing the user to chose which lets the user have more than one clock on the network.

Since the EventIDs are unique, multiple clocks on the network can be configured to coexist. One selectable clock at a time is probably the best option.

In the protocol, one of the clocks is named "Real Time Clock," implying that it runs at a rate of 1:1 and perhaps at the actual time of the location. While this may be a useful clock on a layout, it puts some hard constraints on the implementation. All the issues of real clocks are present, like Time Zones, Daylight Saving Time and time source precision. With the availability of reasonable priced GPS receivers, these issues are now mostly a matter of the correct programming. However, it still is a very complex issue and probably not within the cost for model railroads.

As noted above, time sources are dependent on the type of clock selected. The precision of the time source also depends on how long the clock is running. Requiring precision over months of real time is quite different than several hours. Since most Fast Clocks are only used for hours, the clocks do not have to be very precise. In practice, relying on the µC processor clock is sufficient.

One option which needs to be decided early is how to power the clock. Many installations will rely on the LCC power so that no external power source will be required. Providing power to the network is also a choice that could be made.

The main reason for a Fast Clock on the LCC network is to get time events for use in lighting and clock displays The Clock has to know what Time Events are desired so it can send them on time. LCC Nodes announce what events they consume, using the Consumer Identified message, either when they start or when they are added in configuration. The Fast Clock watches for these messages and keeps a list so that when the time comes, the timer event can be sent. Implementation of this is mandatory. The implementation decisions that remain include how many events, how to manage the list and how to respond to starting mid-way in the list. External observation cannot determine the first two, but the third may be observed. The simplest action is to ignore everything that is prior to the starting time, after all, it will catch up when it cycles through the events. The other option is to attempt to send events that would have been sent prior to the starting time. This is somewhat difficult. A complete cycle through from the starting point back to the starting point would see all of the events, but would be time-consuming. Another option would be to run through the events from 00:00. This will be examined in the next section.

Additional implementation details are left up to the designer. These will be examined for the Logic Rail Fast Clock in the following section.

# The Logic Rail™ LCC Fast Clock

Logic Rail™ has released a LCC Fast Clock which implements the Broadcast Time Protocol. This device comes in two flavors which differ primarily in the size of the display. The device is shown in Figure 9.1. This is the version which has 0.8" digits for a display. The second version has much larger digits – 2.3" – and no additional frame or front buttons. Operationally the two versions are the same. Both come with a power supply independent of LCC and have

**Figure 9.1** **The Basic LogicRail Fast Clock has a Faceplate with a Display and some buttons to Control It.**

options to select the power source. Mounting instructions are included in the

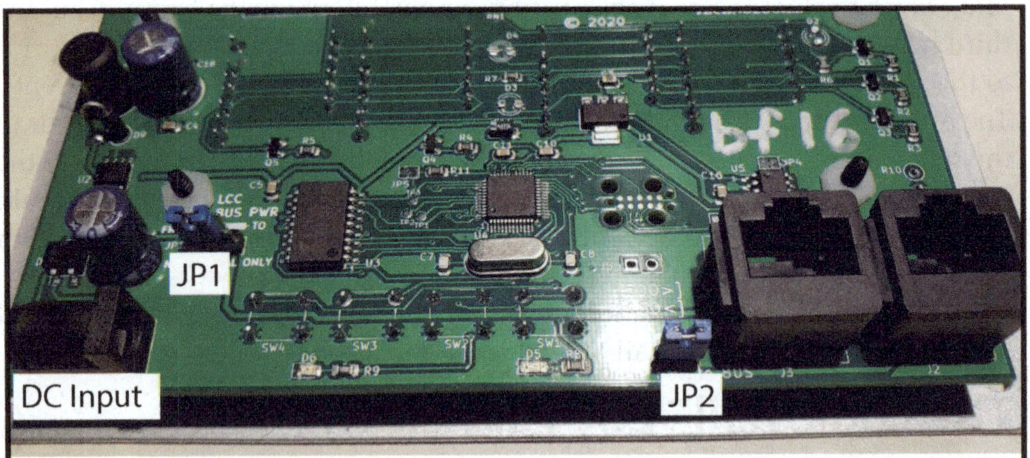


**Figure 9.2** **Power Options for the LCC Fast Clock are set by the Blue Jumpers**

User Manual. This User Manual is quite good, although somewhat difficult to follow. It emphasizes setting up the clock with the front buttons (or back in the Large digit version.)

To avoid multiple button pushes and to provide more flexibility in entering data, let's work with the JMRI configuration. Either way, before you can use the unit, you need to configure the power source option. One of three options is selected by the position of jumpers on the back of the unit. This needs to be set before mounting. The three power options are detailed in Table 9.3. The location of the jumpers is shown in Figure 9.2

**Table 9.3 – LCC Fast Clock Power Options**

| Power Source | JP1 | JP2 |
|---|---|---|
| Bus Power – Power from the LCC Network Cable limits other uses. | Connect Left and Center Pins | Connect Both Pins |
| Self Power – Power from the DC Input for this unit and connects up to 500 ma to the LCC network. | Connect Center and Right pins | Only put Jumper on one pin |
| Local Power – Power from the DC input for just this unit | Only put Jumper on one pin | Connect Both Pins |

Power Options may have an impact on the network by drawing too much power from the network. Displays, especially large ones need a lot of power. According to the User Manual, the clock requires 150 ma. One of the options that supply power through the DC Input would allow the clock to run without affecting the rest of the network. Care should be taken if the "**Self Power**" op-

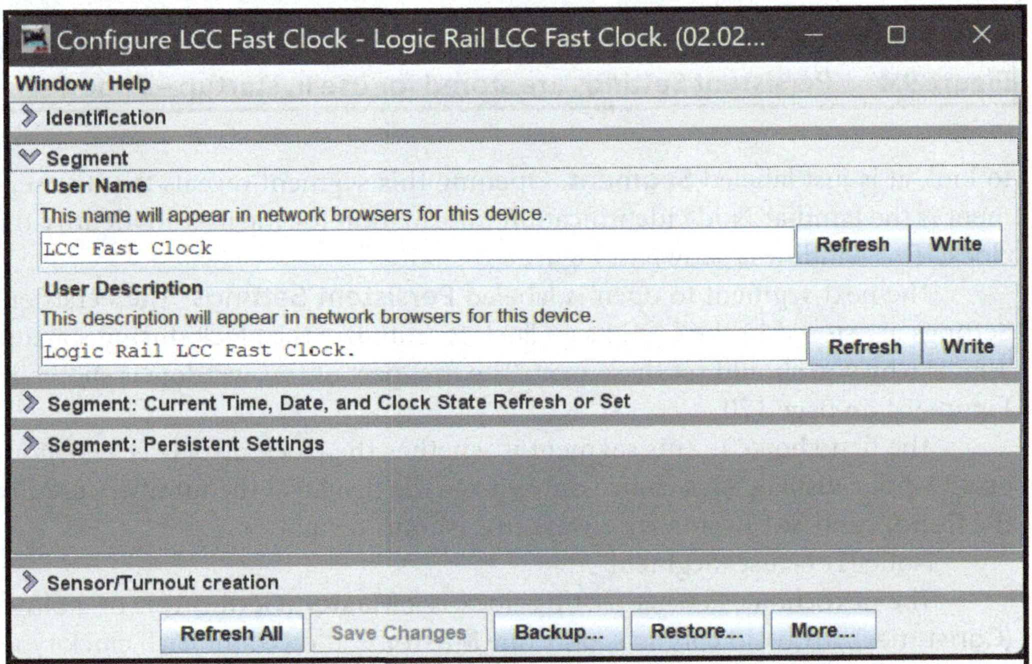


**Figure 9.3 The First Segment is not Named in the LCC Fast Clock**

tion is chosen, because the voltage level on the network should match the voltage used to input the power. Large differences may cause damage to other units. Both the Logic Rail and the RR-CirKits units come with +15V supplies, so they should be compatible.

When the power is correct, the LCC Fast Clock can be connected to the LCC CAN network. Using JMRI, the normal steps will open up a Configuration window. The Configuration window has the normal features along the top and bottom, and, as always, it lists a number of segments to open. The first segment

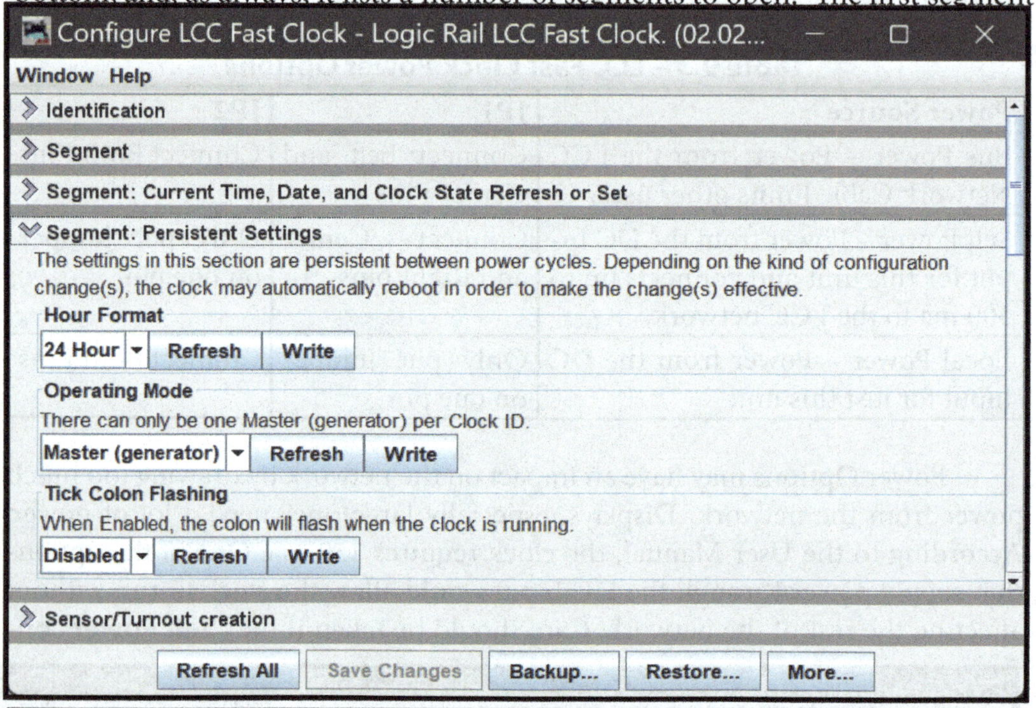


**Figure 9.4 Persistent Settings are stored for use in startup – Page 1**

to look at is just labeled **Segment**. Opening this segment reveals that the segment is the familiar Node Identification information for the user to identify the clock. This window is shown in Figure 9.3

The next segment to open is labeled **Persistent Settings**. The Persistent Settings are stored in the LCC Fast Clock to initialize the clock during startup. That is why you should set these next. The first part of this window is shown in Figure 9.4 on page 170.

The first choice in this segment is whether the clock display is a 24-hour or a 12-hour display. This choice only effects the display – the numbers used in the **Report** and **Set** Events are always in 24-houri format.

Node ID -- just "Segment"

The next choice is whether this clock is a **Master (Generator)** or a **Slave (Consumer)**. To avoid conflicts, only one **Master** is allowed for each clock type.

**Master** clocks are the units that produce time events and respond to **Request** events. The **Master** clock generates time events to synchronize all other clocks using the same clock ID.

If a clock is set to be a **Slave**, none of the other configuration data is important; only the **Master** uses the rest of this data. A **Slave** clock can only de-

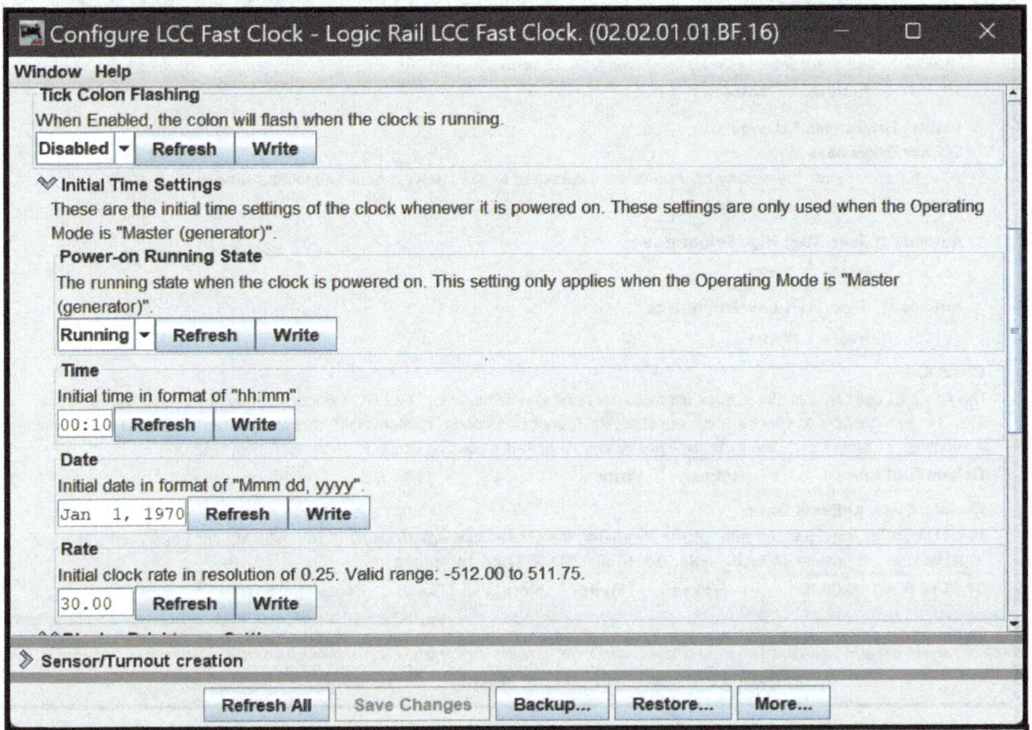


**Figure 9.5 Persistent Settings are stored for use in startup – Page 2**

pend on one **Report Time** Event per hour, so an internal clock must be implemented to provide any intermediate times. To stay synchronized with the **Master**, a **Slave** needs to listen for **Set Rate** events and adjust its internal time keeping to match the **Master**. The Slave also has to respond to **Start** and **Stop** Events to match the **Master** Clock.

The last item on this page is the Tick Colon Flashing setting which will flash the colon between the Hours and Minutes when the clock is running.

The next page of the Persistent Segment sets the initial conditions for the clock. This information is stored and applied on startup or reset. This page is shown in Figure 9.5 on page 171. The settings for Running State, Time, Date and Rate are explained with the format and limits for each.

The bottom page of the persistent settings has some important choices and provides some useful data. Figure 9.6shows this page.

The first item to check on this page is the **Clock ID**. Make sure the correct clock is selected from the pull-down list. When you choose one of the IDs, the next field, labeled **Custom Clock ID Base**, is filled in with the first portion of the

EventIDs for the selected clock, padded with **00.00** to create a full **EventID**. Using **Copy/Paste**, this field can be transferred to other Nodes to control the clock or respond to time events. By editing the last two bytes, any of the Standard Events can be created. If you define your own clock, you have to fill in this

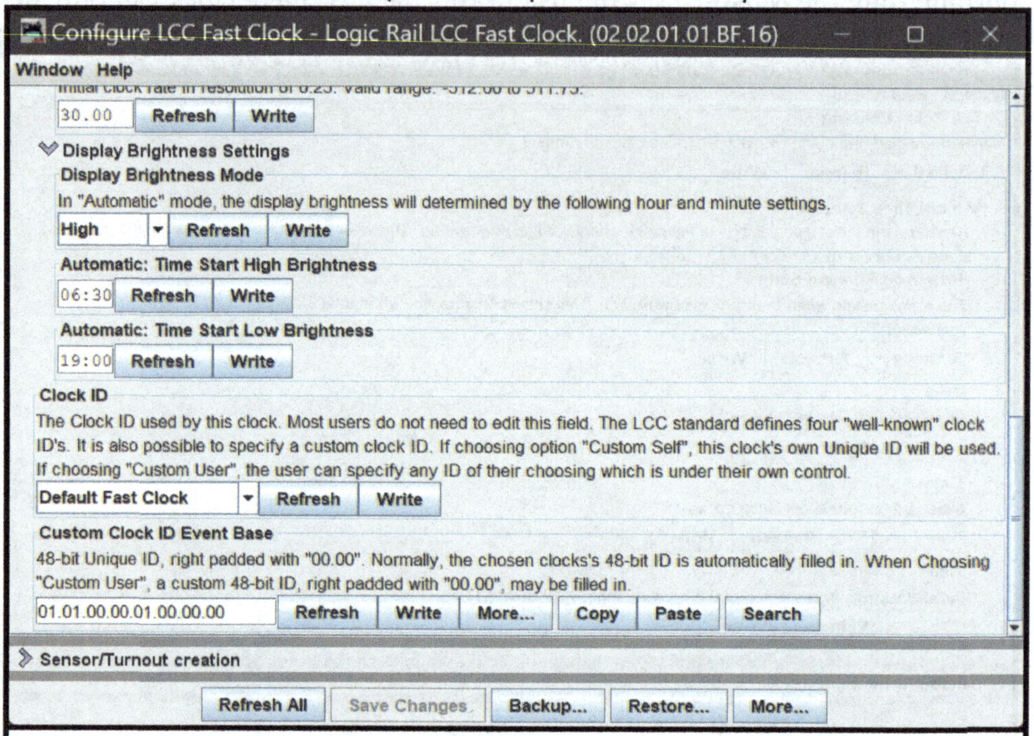


**Figure 9.6 Persistent Settings are stored for use in startup – Page 3**

field for the clock to operate.

The first part of the page in Figure 9.6 sets the brightness of the display. The pull-down list has three options: **High, Low** and **Automatic. High** and **Low** are levels that apply to the clock display continuously. **Automatic** is time related. When **Automatic** is selected, The next two fields are entered to define the times when the clock changes brightness.

These few choices at the bottom of the Persistent Segment are all that are applicable to a Slave clock. All other entries in Configuration are provided by the Master clock during startup.

Going to the next Segment up, labeled **Current Time, ...,** the entries are all Temporary. These entries are filled in when the Configuration is started. The data shown in Figure 9.7 is very temporary, especially when the clock is running. The interesting thing about the **Write** options is that the action is to generate the Set Events to cause the change.

The last item on this page, **Reset Time,** is a little different. If you choose **Enabled**, and then **Write**, it resets the clock to the Initial values in the **Per-**

**sistent segment.** If you click on **Refresh**, it turns its value to **Disabled**. It works as a Reset button; it does not store a value.

In the Standard and in these Configuration segments, the Date is one of the options that can be entered. I am not aware of any displays that include the date, nor do I know of any Operations plans that include multiple dates. The

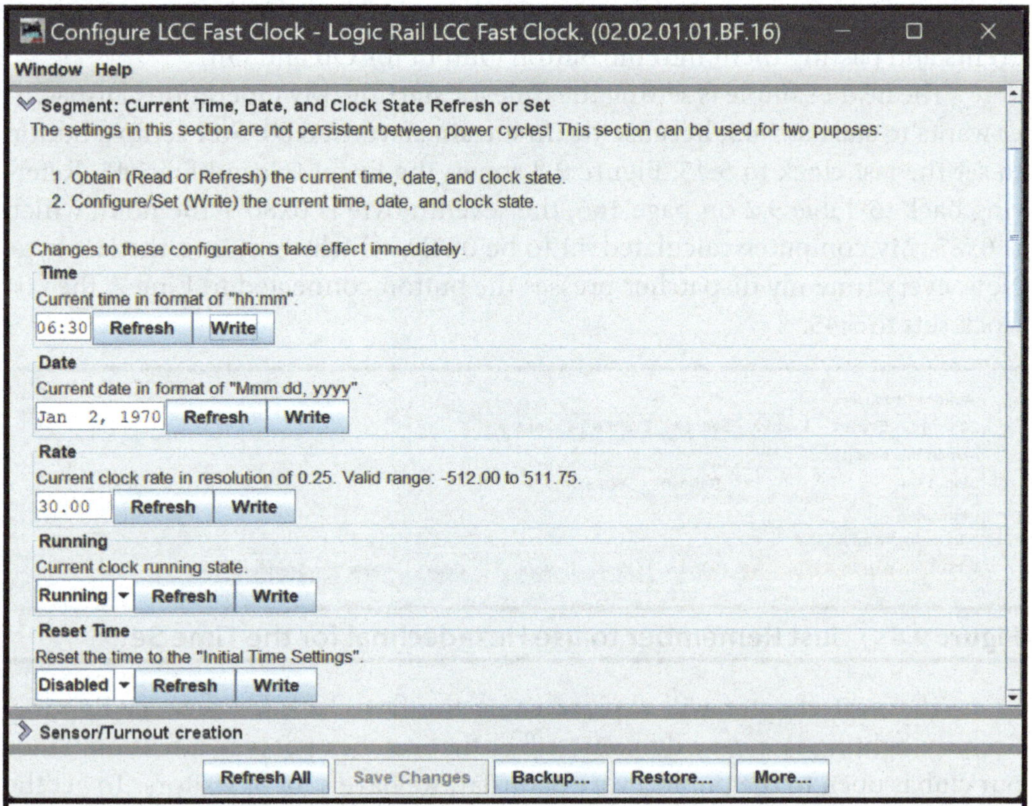


**Figure 9.7 The Current Clock Data is set by this Segment**

option is there to use or ignore.

The JMRI Configuration Tool is one way to provide control of the clock to set it, stop it and start it. Other ways include dedicated buttons on another node or use of the front panel buttons. In the next section dedicated button examples will be presented.

Pausing and stopping the clock from the front panel is only available on the standard size clock. Referring to Figure 9.1 on page 168, the center two buttons provide Pause/Start control. Pressing the left-hand button (Up Arrow) the clock will Start. Pressing the Right-Hand button (Down Arrow) will Pause the clock

## Example Control Buttons for Clock Set Events

To control the fast clock with simple buttons connected to a TowerLCC is simply a matter of configuring the Line in the TowerLCC. Figure 9.8 shows the steps to set up a button. The steps assume that a button is attached to **Line 1.** Using the Alternating form of Input allows a single button to switch back and forth between **Start** and **Stop**. Figure 9.8 only shows the parts of the configuration segment that need to be entered.

If a Sampled pushbutton is used, an added feature would be to light the button when the clock is running. This addition is left as an exercise for the reader. Hint: it only requires changing the output to Steady then copying two events and pasting them into the button Output for On and Off.

The next example is setting the time for starting. My Operations Dispatcher wants to start at 5:45, because the first train starts at 6:00. So I setup a button to set the fast clock to 5:45. Figure 9.8 shows the Event to send for 5:45. Referring back to Table 9.2 on page 166, the seventh byte is 0x80 + the hour, which is 0x85. My computer calculated 45 to be 0x2D, which went into the last byte. Now, every time my dispatcher presses the button connected to Line 2, the fast clock sets to 5:45.

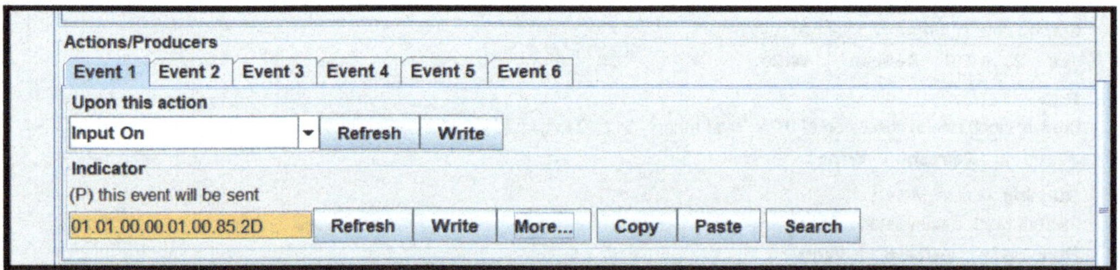


**Figure 9.8 Just Remember to use Hexadecimal for the Time Set**

The next chapter will explore using the fast clock for scenery lighting. Scenery lighting is a very dynamic effect that catches people's attention. Since our club is open to the public every Saturday, we have a lot of visitors. To get the time-based lighting on and off within the time of a visit, a very fast rate can be used. I set that rate at 60 so a full day cycle only takes 24 minutes. The only

**Figure 9.9 Set the Clock Rate to 60 for Display Purposes**

problem is that the public does not understand, so I posted a note beside a fast clock slave display that says " *In the world of the Belmont Shore Railroad, time moves quickly. Because the railroad is built to a scale of 160:1, distances are shorter than in the real world. In addition, some distances are "selectively com-*

*pressed" to bring towns and cities closer together. Because of these factors, we operate on a faster clock than normal*

*During visiting times, we run our clock even faster to display the changes in lighting through a 24 hour period."*

# Fast Clock Slave Displays

In a large layout or just for interest, slave displays can be added. Slave displays as mentioned earlier need to listen to the LCC network and react to changes so that the time is synchronized with the **Master**. Three types of Slave clock are discussed below.

## Using the Logic Rail Fast Clock

The easiest Slave clock is probably the Logic Rail Fast Clock set to Slave mode. This may be the most expensive way to get a Slave Fast Clock, but it is a reliable device with no soldering or assembly involved. Use as a Slave is overkill. Much of the expensive design is unused. It is simple to install – simply plug it into the LCC network and set the configuration to be a Slave.

## JMRI Display Slave Clock

JMRI has several clock options, including LCC. JMRI even has a Master Fast Time option. Unfortunately, the JMRI "Master" does not generate any of the Time Events necessary to be a Master. This implementation is currently incomplete – it may be completed later, but at this writing it is not a true Master.

As a Slave, JMRI has some excellent uses. For example, it can be displayed on a large screen., Or multiple screens. Setting it up requires several steps:

1. From PanelPro Main window, select Edit
2. In Edit window, select Preferences
3. In Preferences window, select LCC Protocol
4. In Protocol, select Fast Clock Tab
5. In Protocol, select Clock ID and Clock Slave /Consumer( if necessary
6. Save to get back to Preferences
7. In Preferences, select Save, then choose Restart
8. Wait –when JMRI restarts, select Tools
9. In Tools pull-down, select Clocks and Setup Fast Clock
10. In Setup Fast Clock, select Time Source – OpenLCB …

**Figure 9.10 JMRI Displays Clocks In Different Styles**

11. In Fast Clocks, select Done
12. Back in JMRI main window, select Tools
13. In Tools pull-down, select Clocks
14. In Clocks List select the style of clock

That should give you a clock display synchronized with the LCC Fast Clock. Two of the styles are shown in Figure 9.10 on page 175.

### DIY Slave Clock – The Lurker

Since we know the Events that control and are reported by the LCC Fast Clock, a DIY Slave Clock is a possibility. All it takes is a μC with a CAN I/O. Fortunately, there are several available, including several which can be programed with the Arduino IDE. One of these, the Adafruit Feather M4 CAN has the correct I/O and interfaces to listen (and talk, although that is not necessary for this application) to the LCC CAN network.. The Slave Clock shown in Figure 9.11 is not formally a Node on the LCC network – it is a lurker observing the network but not accessible from the network. It simply listens to the messages on the network, acting on those that it needs.

**Figure 9.11 DIY Slave Clocks can use Optional Displays**

A DIY clock has the advantage that the display can be different than the available product. Scale size displays could be placed on buildings. As a future project, the Time Events could be sent to a WiFi or Bluetooth transmitter and then sent to a pocket watch, a real engineer's watch running on Fast Time.

How does the Lurker work? It listens to the messages on the LCC network and responds to those it requires to keep accurate time. Fortunately, Adafruit provides a library that reads CAN messages and sends the data portion to the program. The data portion is the Events that are generated by the Fast Clock. Only 4 event types are required — **Time**, **Rate**, **Start** and **Stop**. Each is processed and the result stored in a variable. The only one that takes very much processing is **Rate** (see the discussion of the **Rate** fields above.) Then, if the clock is running, the current time is updated and displayed.

Since the Fast Clock does not send a **Time Event** very frequently, the Lurker has to maintain its own internal time. It does this using the Arduino function (millis) which returns the number of elapsed milliseconds since start-up. The **Rate Event** is converted into the number of milliseconds between clock increments to provide the millisecond number for the next output update. One of the features of the Fast Clock is that the **Start Event** always sends along the **Time Event** and the **Rate Event**, so **Start** updates everything for internal time-keeping.

The Lurker is an example of using the LCC data without all the initialization necessary for a full node. It does not have everything to be recognized on the network, and cannot truly be considered an LCC node. For more details see Appendix B.

# Summary

LCC Fast Clocks are enabled by the Broadcast Time Protocol. The Events defined by the Standard provide a basis for Fast Clock implementation. The Logic Rail implementation was examined, including the setup and control. Applications to control scenery lighting were mentioned, but details have been left to the next chapter. Clock displays are also described, including commercially available slave Displays, JMRI and a DIY display nicknamed the Lurker.

# Chapter 10 — Scheduling Lights and Other Things

The most impressive scenery lighting adds animation to the layout. This is achieved by turning on and off lights at scheduled times. In real life, lights are dependent on the time and the application. Some lights are only on during the day (e. g. Retail stores and office buildings) and others are only on at night *e.g. street lights and residences.) Exceptions can occur, but either having all lights on all the time or turning all lights on at the same time is not realistic. Using a Fast Clock as described in the previous chapter can add realism and action to a layout.

Fast clocks are an excellent choice for scheduling scenery lighting, because the observers see a full days worth of light changes in a short time. The rate of the Fast Clock depends on the observer. During Operations, train crews are normally aware of the simulated time of day, so they expect street lights at night, roundhouse lights both day and night, and stores in the daytime. Other visitors are not likely to understand what time the fast clock simulating and they are not likely to stand watching the lights change if they are not informed about fast clocks.

I would suggest a simple sign to inform the visitors about the fast clock. For the Belmont Shore club, I propose a Fast Clock Display titled “Belmont Standard Time” and this explanation:

> *"In the world of the Belmont Shore Railroad, time moves quickly. Because the railroad is built to a scale of 160:1, distances are shorter than in the real world. In addition, some distances are "selectively compressed" to bring towns and cities closer together. Because of these factors, we operate on a faster clock than normal*
>
> *During visiting times, we run our clock even faster to display the changes in lighting through a 24 hour period."*

In addition, I would suggest a very fast rate, maybe 60:1. This would make a full day in 24 minutes, which would display most changes in about half that time. As a visitor looks around, lights will be on and off to emphasize the scenery.

### A Brief Review of LEDs

In Chapter 2 LED use was introduced. Now, for scenery lighting, multiple LEDs will be everywhere. The temptation will be to put many LEDs on one wire to save time and effort. In most cases this is not a good plan. Remembering that the LED is a constant voltage device, too many in series require higher voltage (the sum of all the voltages on the LEDs) and can cause the "Christmas Light" effect where one breaks and causes the whole string to go out. For that reason, this should be limited to no more than three or four.

Connecting LEDs in parallel on a single resistor is another temptation. It is not to save cost, since resistors are very cheap. Maybe it saves time, but the problem of varying LED voltage causes some LEDs to be brighter than others and can lead to excessive current on LEDs if one or more go out.

In most cases, a single LED connected to a single resistor is the best choice. Some exceptions are possible, as shown later in this chapter.

## Start with a TowerLCC

To understand how to use Time Events, start with a TowerLCC. Previ0ous chapters have used the TowerLCC to turn on LEDs, so this is not much different. The main concern is getting the correct events for the time desired. There are some limitations when using the TowerLCC because it is not designed for adjusting the brightness of an LED. Therefore, additional parts may be needed for a permanent use. Availability of PWM module makes it easy to provide a wide range of brightness, so that is easily fixed.

When considering what can be controlled by the TowerLCC, remember that the TowerLCC, just like the Arduino modules, has GPIO pins connected to the µC. Therefore, if a device claims to be Arduino compatible, and it uses GPIO pins, it is compatible with the TowerLCC. This compatibility opens up a wide range of specialty modules, including sound and other special effects. The only limitation is the current capacity of the TowerLCC. Methods to increase the current output will be presented later in this chapter.

Even with the limitations, the TowerLCC is a good module to use because most layouts will have a line or two available on TowerLCC modules. Lighting

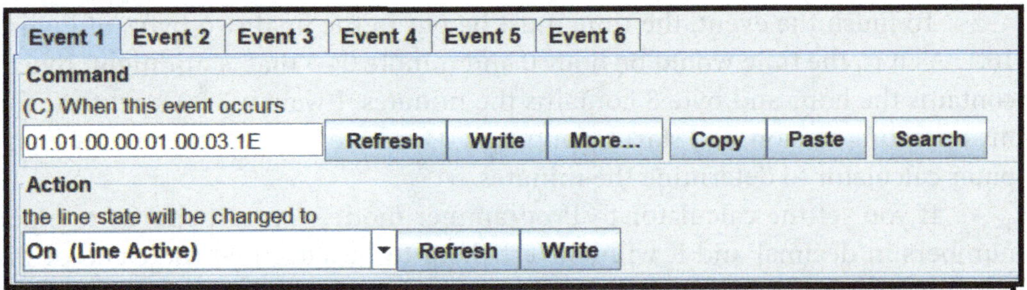


**Figure 10.1** Clock Events for the TowerLCC need to be created.

and other time-based effects will be spread all over the layout, so sharing modules is a good way to use otherwise unused lines.

## Creating Time Events

The Broadcast Time Protocol Events shown in Table 9.1 on page 164 define what the event looks like. It consists of two parts, the specific Clock's ID and the Time in Hexadecimal. The Clock ID is available from the LCC Configuration window, segment labeled **Persistent Settings.** At the bottom of this segment is a section titled **Clock ID.** As you saw in the last chapter, this is where you define the specific Clock to use. Below that section is the **Custom Clock ID Base,** which has an entry containing a model of the events for that clock. Contrary to its title, it contains the model for the standard clocks. In the case of the **Default Fast Clock,** this model is **01.01.00.00.01.00.00.00.** That is the start for the EventID that we need to use. Paste

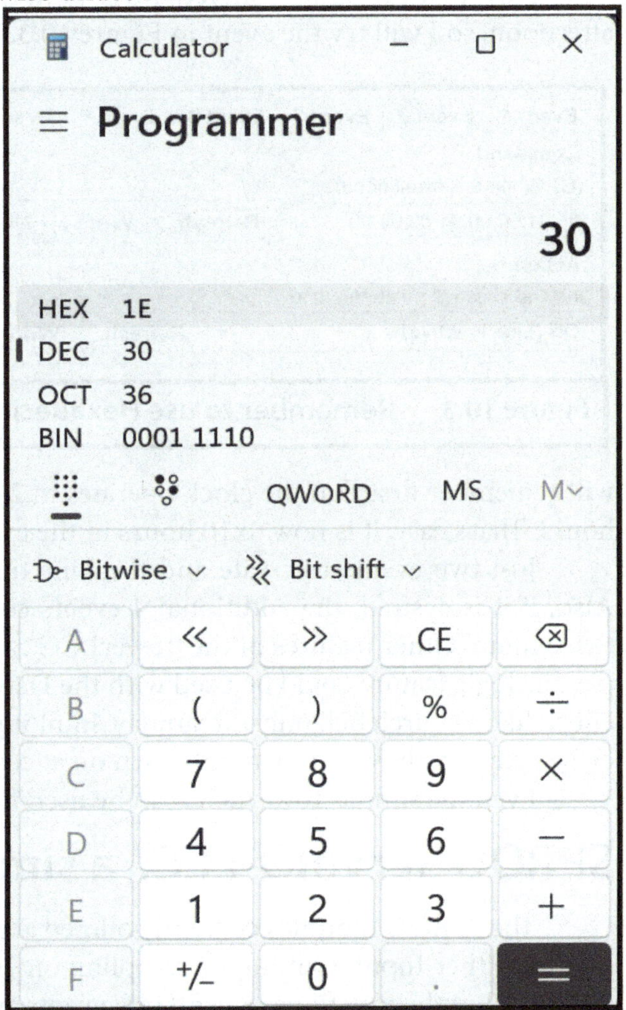


**Figure 10.2** The Computer Programmer Calculator is great to get Hexadecimal

that model into the Consumer Event in the TowerLCC, like Figure 10.1 on page 181.

To finish the event, the time must be put in the final two bytes of Figure 10.1. As it is, the time would be hour 0 and minute 0 — that is midnight. Byte 7 contains the hour and byte 8 contains the minutes. I want a 3:30 start time for my operating session, so I start by putting a 03 in byte 7. Now I turn to my computer calculator to determine the minutes.

If you set the calculator to Programmer mode, you can choose to enter numbers in decimal and it will convert them to hexadecimal (or Octal or binary). Using the calculator avoids a lot of head-scratching and you don't have to take off-your shoes to count. This calculator is shown in Figure 10.2 on page 181. The input number, 30, is 1E in hexadecimal. (Programmers would make that a 0x1E to indicate it is a hexadecimal number.)

Now that the light is on, let's turn it off. I want to turn it off at 4:00 in the afternoon, so I will try the event in Figure 10.3. When it turns off prematurely, I

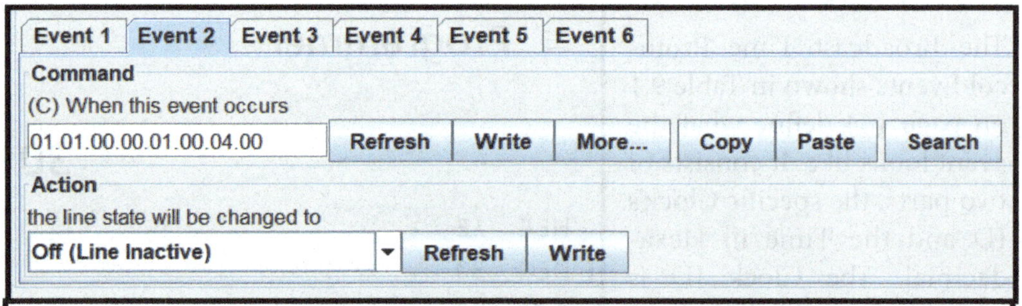


**Figure 10.3 Remember to use Hexadecimal in the Events**

will remember first that the clock operates in 24 hour time, so 4:00 PM is 1600 hours. That's easy, it is now 0x10 hours in the event.

Just two events to create and the Line turns on at 3:30 and off at 16:00. Also, if it is desired, the additional 4 events could be used to turn on and off twice more. Other features of the TowerLCC can add to the realism. For example, the logic feature could be used with the Date events to provide a day related effect, like a church bell only on Sunday. Blinking could be used for an emergency vehicle, which would then only turn on at certain times.

Even the simple TowerLCC can be used for time-related effects.

# SPROG ServoIO-LCC, a similar Start

The SPROG modules currently offered are similar to the TowerLCC. They provide either Input, Output, or sampling on 16 lines. The connectors may be different, but the functions to use Time events are essentially the same. The one difference that is useful is that these modules have a slightly higher current capability — 50 ma instead of 25.

# Logic Rail EFX-16 — special effects

Building on the Fast Clock, Logic Rail produces a special effects board to perform lighting. The EFX-16 board connects to the LCC network to receive the Time Events generated by the Broadcast Time Protocol. It has the ability to receive events from any of the defined Fast Clocks. It also can interface to the DCC signal, which I will not detail. In Figure 10.4, the connectors for LCC are on the

**Figure 10.4 The EFX-16 connects LCC to LEDs for Scheduled Lighting Effects.**

left edge, with an Auxiliary DC input next to them. Along the top and bottom of this photo are the connectors for lights. The EFX-16 uses the same connectors as the Woodland Scenics JustPlug™ Lights. Prewired mating plugs (JST 2.54)are available from Amazon and other sources.

On the right side is a connector for a special Button Board which can control the lights when installed. Also on the right end is a 2-pin screw terminal labeled MST ON/OFF. If the Button Board is not installed, this can be used to turn on or off all of the lines on the top and bottom suds.

The connectors on the top and bottom are normally connected to "Constant Current" drivers on the board. These drivers attempt to provide a constant current of 20 ma. by varying the voltage until either that current is achieved or the maximum voltage available is output. This is convenient for driving one to three or four LEDs. Since most LEDs are rated for 20 ma. current, this ensures that an LED cannot be overdriven, and that no resistor is required. Brightness is controlled by setting the period of internal PWM on each line.

If more current is needed, A modified board can be ordered that removes the constant current driver and sends the full input voltage to the connector. In Figure 10.4, the connectors with yellow tape on them have been modified by Logic Rail. These lines still perform all the special effects, but now can source up to 200ma. Proper resistors must be connected to prevent over driving LEDs connected to these lines.

## Initial EFX-16 Hardware Setup

In Figure 10.4 there are two blue jumpers. These jumpers control the power selection for the EFX-16. These option settings are defined in Table 10.1. Jumpers that are "Only One Pin" are stated that way to keep the jumper with the board. Otherwise the jumper connects the identified pins. The primary decision is whether to take power from the LCC bus, provide power to LCC, or neither. If the EFX-16 is powered from the LCC Bus, no added power supply is needed. Otherwise, a power source must be provided.

**Table 10.1 – Power Options for the EFX-16**

| Power Method | JP1 Jumper | JP2 Jumper |
|---|---|---|
| Bus Power (from LCC) | Top & Center Pins | Both Pins |
| Power to LCC Bus | Center & Bottom Pins | Only One Pin |
| Aux Power Only | Only One Pin | Both Pins |

One other situation will require an external power source. If the board is modified for higher power, that probably means the LCC bus will be overloaded. In this case, the last two options would be preferred.

When power is now provided, things will light up. Most importantly, the green LED below the EFX-16 logo will light indicating that power is on. If you have LEDs connected,, you may find that all lines are On. If that is what you want at start-up, no problem. If not, no problem, I will show how to change this later.

## Initial Configuration Setup

A few things need to be done in Configuration before the lights are configured. Primarily the standard items should be filled in and set. Open the Configuration Tool and open up the EFX-16 Configuration. Open the first Segment titled **Device Settings.** In Figure 10.5 on page 185 is the familiar identification to help you in getting the correct board when configuring a large number of nodes. Fill in the fields, save them and open up the segment titled **Lights, Inputs and Effects.** This is the primary segment that will be used to configure everything. Most of this segment is related to individual lines, but way at the bottom are the first items to configure.

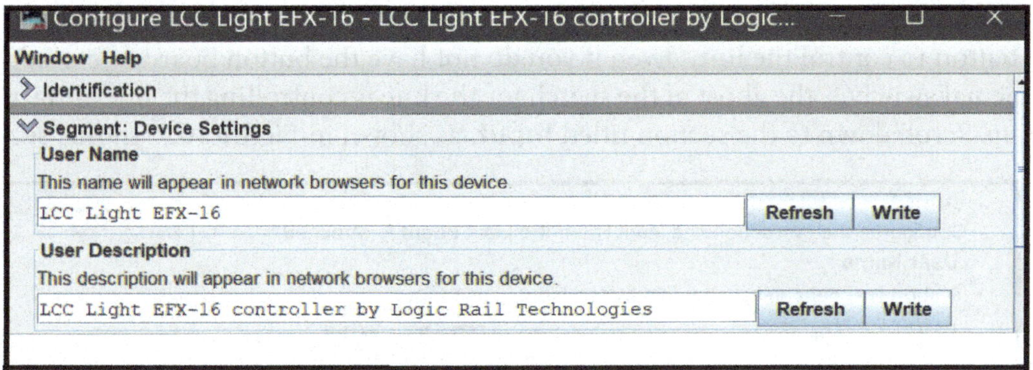


**Figure 10.5 Enter the Familiar Identification for this Board**

The section titled **LCC Configuration** is where the EFX-16 sets the Fast Clock ID. You should also notice that the model event is also repeated here for your use. You will not need it very much, because this device creates its own.

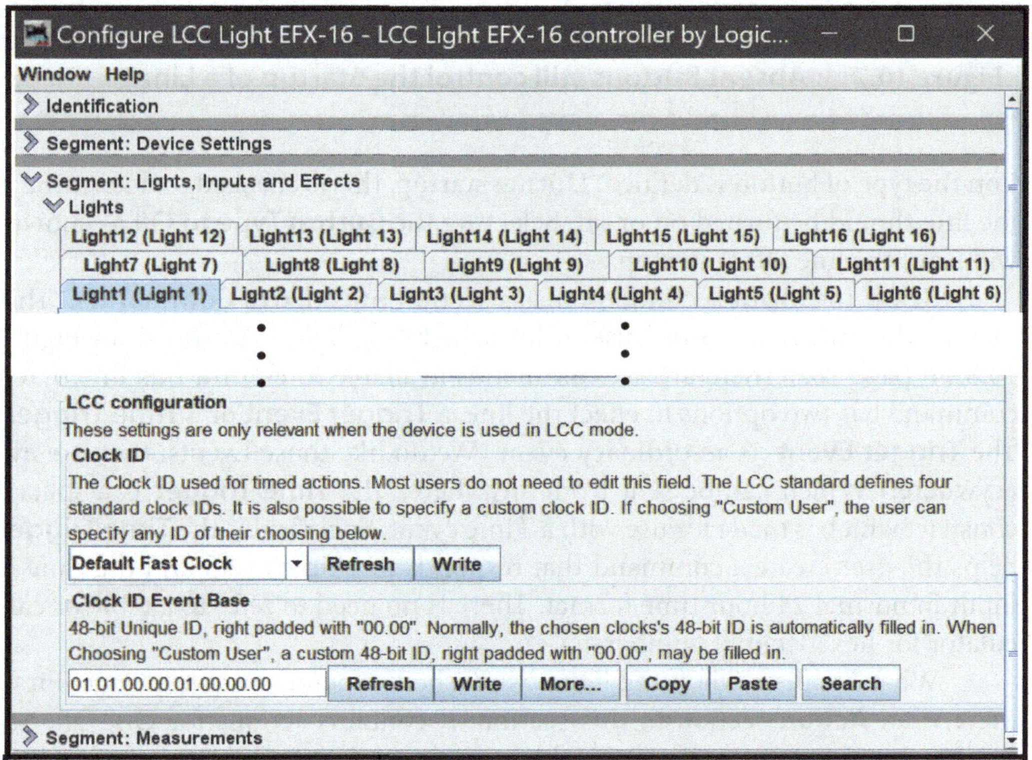


**Figure 10.6 Connect to the correct Fast Clock for Time Events**

Preparing to enter consumers for the various lines includes deciding what the line should be at startup. If the initial state of the line is On, and if that is what it should be, then nothing needs to be changed. If the initial state is On, and it should be Off, then a fix it as needed. The fix is somewhat obscure but it works.

Start by selecting the line to fix as shown on the top of Figure 10.6. This selection provides all the options for this line, including the use of an add-on button to control the line. Even if you do not have the button board to connect actual switches, the ghost of the switch for the line is controlling the line at start-up. Scroll down to the section titled **Input,** as shown in Figure 10.7. In this sec-

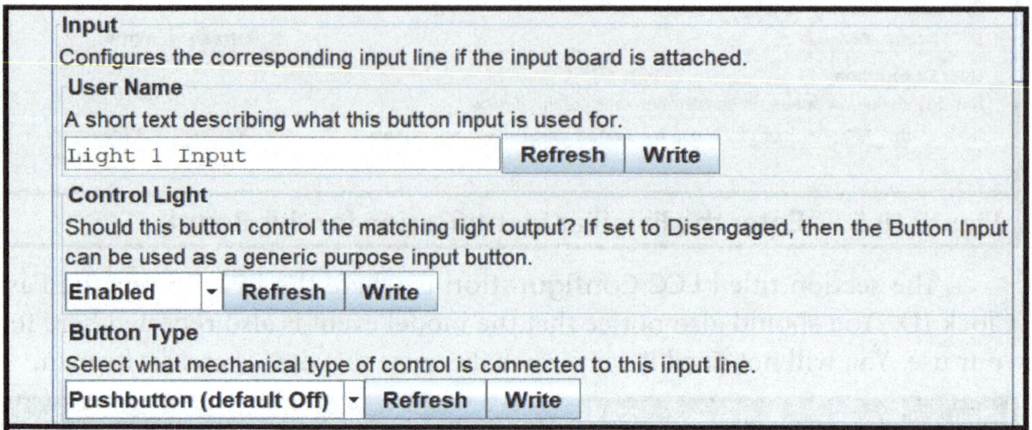


**Figure 10.7 Absent Buttons still control the Startup of a Line**

tion the type of button is defined. During startup, this is checked to determine if the line should be turned on or off. Selecting the **Button Type** to Off as shown, will leave the line Off at startup.

Below the **Input** section, there is a section titled **Extra Commands.** This may be the area that is most useful for scheduling lights. As shown in Figure 10.8 on page 187, There are six commands available to control this line. Each command has two options to effect the line, a **Trigger Event** or a **Time Trigger**. The **Trigger Event** is an ordinary event (We do like those Events, they are everywhere!) Which can be sent from anywhere. The **Time Trigger** is a special constrict which is made for use with a Time event. Specifically, the **Time Trigger** helps the user create a command that responds to a time event and the time is input in normal 24 hour time format. There is no need to refer to a table or calculator for hexadecimal numbers.

What happens when the Time is reached depends on two things. First, there is an **Action** section in this section. It contains actions for On, Off, Set Brightness (Yes you can dim or brighten a light while it is On), Fade in and out, Blink, and some specialized actions for crossing signals which I will not go into. With the exception of On, Off and Set Brightness, these are the effect on the light until it is changed. The first three are applied to the light **Effect** entered in the overall line section. This will be described later.

There is one more entry that may be needed. For Brightness and Fade, there is a parameter field that needs to be entered to get the desired effect.

These **Extra Commands** provide a flexibility in lighting effects because

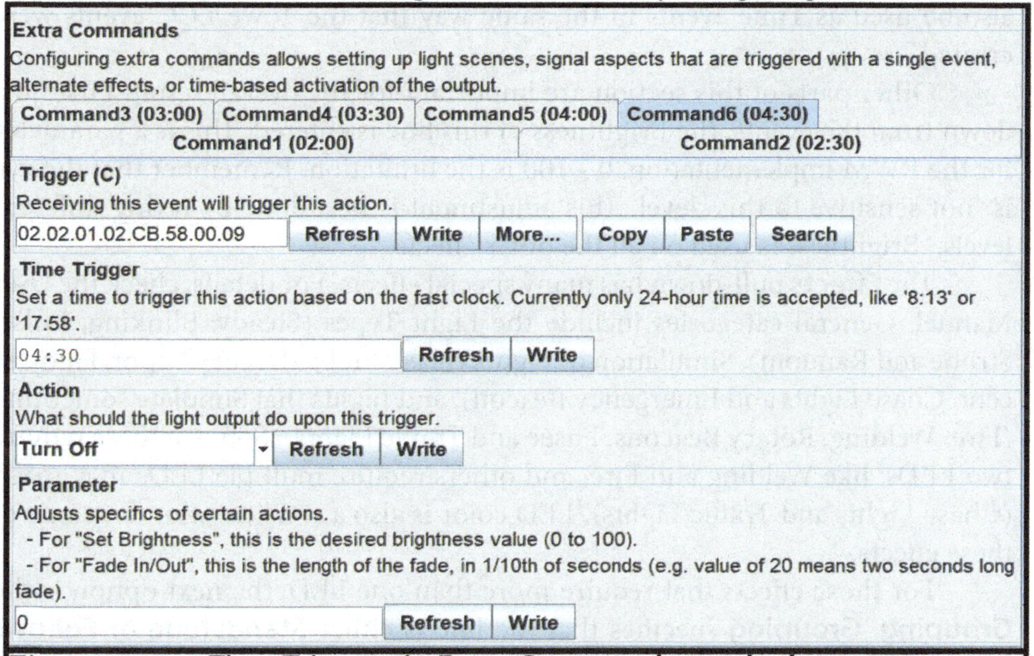


**Figure 10.8 Time Triggers in Extra Commands are the best way to schedule Lights**

they can modify a light at specific times. For example, a police car could be set to turn on its headlights, and then make them flashing for an emergency.

Going back to the top, there are some things in Figure 10.9 that are used for the line. Primarily, there are two events that can be used for On and Off. If

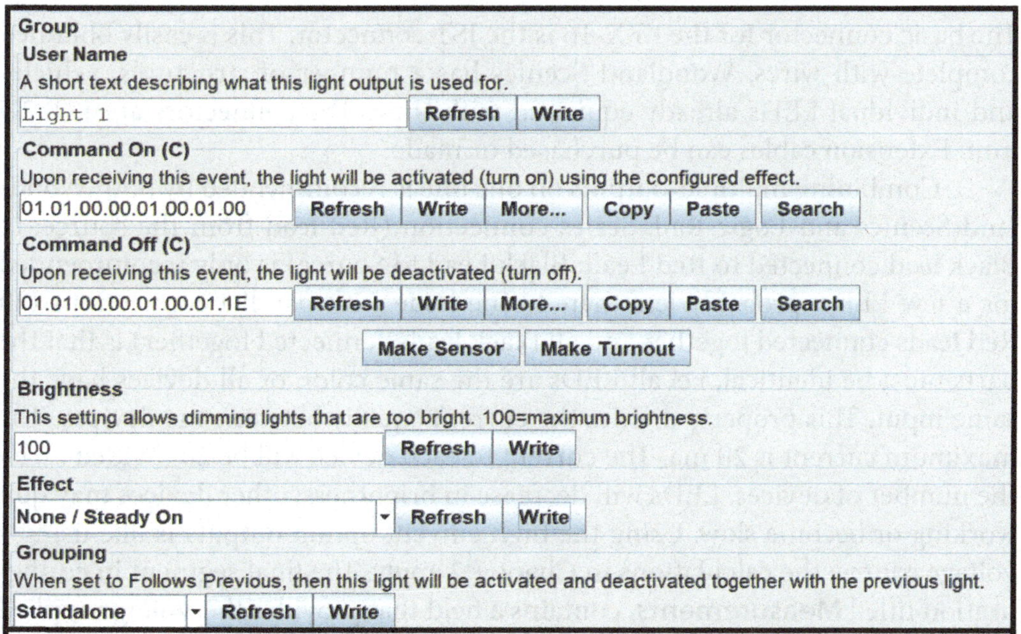


**Figure 10.9 Basic I/O uses ordinary Events for On and OFF**

there is a need for a global layout Off command this is a place to put it. These can also be used as Time events in the same way that the TowerLCC events were created.

Other parts of this section are important for the overall result. First, just down from the events, the Brightness of this line is entered. This is a parameter for the PWM implementation. 0 - 100 is the limitation. Remember that the eye is not sensitive to this level. This adjustment is best done by trying different levels. Brightness is used on all the other effects.

The **Effects** pull-down has many special effects. For details, check the User Manuel. General categories include the Light Types (Steady, Blinking, Pulse, Strobe and Random), Simulation of Lights (Flash Bulb, Mercury Vapor, Fluorescent, Chase Lights and Emergency Beacon), and Lights that Simulate something (Fire, Welding, Rotary Beacons, Fusee and Traffic Lights). Some of these require two LEDs, like Welding and Fire, and others require multiple LEDs in a group (Chase Lights and Traffic Lights). LED color is also a consideration for some of these effects.

For those effects that require more than one LED, the next option is the **Grouping**. **Grouping** specifies that the line is either **Standalone** or **Follows Previous. Follows Previous** should be used for those effects that require more than one line. Using it means that this line is activated at the same time as the previous line.

Additional sections of the Configuration can be explored by the reader. The sections already described are probably the most important for use with LCC.

## Output Connections

The basic connector for the EFX-16 is the JST connector. This is easily obtained complete with wires. Woodland Scenics has a number of structures, vehicles and individual LEDs already equipped with these The connectors are JST 2.5 mm. Extension cables can be purchased or made.

Combining multiple outputs on one line is recommended by both Woodland Scenics and Logic Rail. Series connection (Red lead from the source; to Black lead connected to Red Lead; Black Lead to Source) is only recommended for a few LEDs, probably less than 4. The issue with parallel connections (all Red leads connected together and all Black leads connected together) is that the parts must be identical, i.e. all LEDs are the same color, or all devices have the same input. This properly divides the current equally between the devices. The maximum current is 20 ma. The current to each device will be 20 divided by the number of devices. LEDs will decrease in brightness, other devices may quit working or become slow. Using the high current option outputs is like using a voltage source; the calculations in Chapter 2 apply. The final segment in configuration titled **Measurements,** contains a field that provides the Voltage applied to LEDs. This value can be used in the calculations.

# Woodland Scenics® JustPlug™ Control

A popular source of lighting is the JustPlug line from Woodland Scenics. The JustPlug system includes end items like street lights, structures and vehicles as well as infrastructure like individual LEDs, hubs and power supplies. A starter system would consist of a power supply, a JustPlug Light Hub, and 1 to 4 end items. To expand, an Expansion Hub would be added to split off one to four Light Hubs. As an add-on, an Auxiliary Switch can be purchased to plug into either the Light Hub or the Expansion Hub. To use the switch, a shorting jumper that is plugged into the hub is replaced by the switch .

To control the JustPlug hub, LCC needs to mimic the action of that switch. A quick check determined that the jumper was supplying the positive side of the power. What that means is that the switch is what is known as a "high-side switch." A quick search found a solid-state relay, the CPC1706A. A quick PCB design (after a breadboard test) resulted in a simple PCB that can be connected to either the EFX-16 or a TowerLCC.

> **Printed Circuit Boards**
>
> In several chapters I have shown custom printed circuit boards (PCBs). I have created these boards for this book and for the Belmont Shore Lines. PCBs are ubiquitous in modern electronics (well, starting in the 1960s or 1970s) so why not? I am here to tell you that PCBs are easily in reach of the common model railroader. Many videos on the Internet describe the process and there are several inexpensive or free programs for design. The boards are cheap (none of the ones I have shown are more than $5. Don't hesitate, the advantages are great.

The schematic and board layout in Figure 10.10 shows how simple this design is. The Solid-State Relay is optically isolated which is perfect for this use. An added resistor ensures that the TowerLCC will not be damaged when used in this case. When used with the EFX-16, one or more could be connected in parallel. The switch is supposed to activate at less than 5 ma, however, low currents may take longer to switch. Considering the time to activate is important to maintain a good PWM. The PWM from the EFX-16 is operating at about 200 Hz, which means pulses are 5 msec apart. Since the CPC1706 switches in micro-

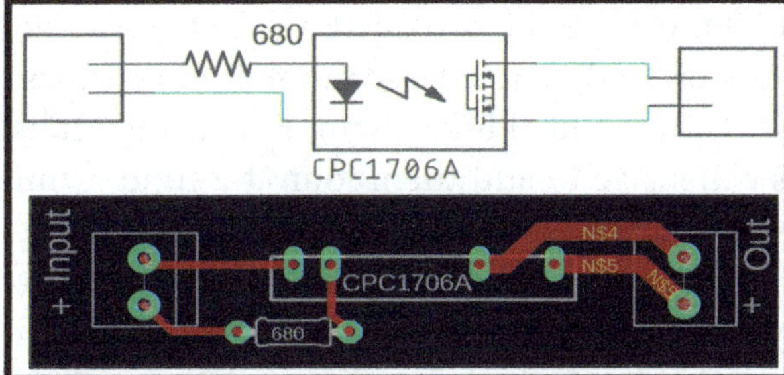


**Figure 10.10 A High-side Rerlay Controls the JustPlug Lights**

seconds. When connected to an EFX-16, this means that all the effects designed into that board can be set from LCC as shown in Figure 10.11

**Figure 10.11 The Solid-State Relay is Connected to Provide Time Event Control for JustPlug Lighting**

Connecting the EFX-16 to either the Light Hub as shown or the Expansion Hub is simple. Prepare the Solid-State Relay by connecting JST cables to the screw terminals on the board, connecting Red wires to the _ side. Plug in the Input side to the EFX-16. Pull out the shorting plug from the port labeled CTRL and plug in the cable attached to the Solid-State relay board. That's it. Now you have the advantages of JustPlug combined with the control of LCC..

# Increasing Current with Add-ons

As mentioned earlier, sometimes many lights need to be switched. More than 4 or 5 will probably overload the previous control methods. Add-on devices can support higher currents for those cases. Two methods have already been mentioned — the modified EFX-16 and the JustPlug relay. In the EFX-16, up to 200 ma. can be provided. Using this method, each LED should have its own resistor, leading to a large number of resistors to hang on LEDs.

In the JustPlug system, Expansion Hubs can add many LEDs, although it is expensive to add them four at a time. Connecting two or three in each plug will cut down the total number, but will reduce the brightness available.

A traditional method to control high current devices is the mechanical relay. Standard relays are often used for control of motors because of the back current due to the impedance of the motor. Both RR-CirKits and Logic Rail have optional relays for their produces. The problem with relays is that they are slow and rely on moving a set of contacts to work. This makes them impractical for PWM or other situations where fast changes in the output are necessary.

Another consideration for lighting is why is there a requirement for higher current The most common reason is the total number of LEDs that are grouped together. An example is shown in Figure 10.12. In this case, approximately 100 LEDs are spaced along the streets in this city. Each one is set to 10 ma. for initial levels. Brightness will be adjusted after installation. This means that not only will there be a significant current, there will be a lot of resistors. Working under the layout can be minimized by a custom PCB to support the street lamps. The Board in Figure 10.13 supplies screw terminal connections for 20 street lamps at a time. Considering that the streets are spread over an area about 8 feet long and 3 feet wide, distributed support makes the connections shorter. Five boards will be connected to one EFX-16 line, which means that each one should be provided 4 ma.

**Figure 10.12 Many Street Lamps add up to High Current**

This same street scene will have traffic signals operated by the EFX-16. It is anticipated that at least 10 intersections will have traffic lights. This adds up to another 400 ma (2 LEDs at a time, in each direction is 4 LEDs per intersection). but only 200 ma. maximum per line. The plan is to have a small PCB at each intersection which has the 6 LED connections and the 6 resistors. A ribbon cable will connect these back to the EFX-16 interface which will contain the solids-state relays.

**Figure 10.13 Relay Board to Connect 20 Lamps at a Time**

These examples show that the control of lights is very possible with the devices available. Commercially available devices like the JustPlug system can be easily merged into the LCC for scheduled time control. Now let's look at some other things to control.

# Other Things to Schedule

Once we have time events, lights are not the only things that can be scheduled. Anything that can be triggered by an Arduino can be triggered by LCC. A quick look at some things will start you on this exciting exploration.

### Sound

Looking at the sound modules sometimes found in greeting cards, you might wonder why that couldn't be connected to LCC. Well, of course it can. A search on the Internet turns up some sound modules which come with a speaker and a button to start. I believe if it has a button, I can trigger it from LCC. The module shown in Figure 10.14 provides 8 Mb of MP3 sound. It is loaded from a Windows PC and powered by 5 volts, maybe from a TowerLCC The best thing about this is that it can do most any sound, from an opera singing in the city, to church bells, to cows mooing in the pasture and a foghorn in the harbor.

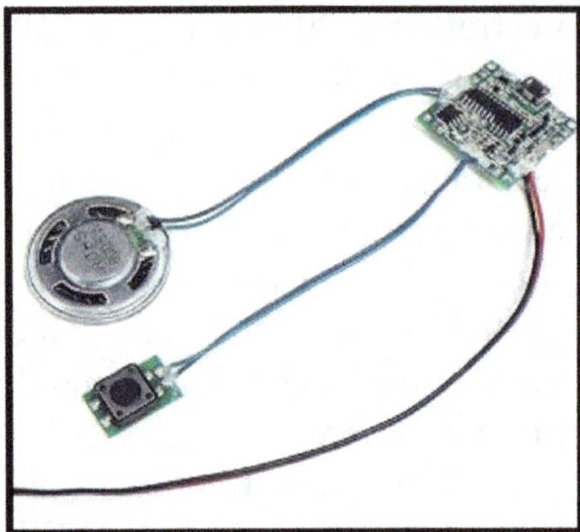

**Figure 10.14 A Mini Sound Module for scheduled Sounds**

Search for "sound module" and you will probably turn up many to choose from.

### Animation

Every layout needs some additional animation. Of course all model railroads are loaded with trains, but additional animation shows up in amusement parks, sawmills, mines, and other things. If scheduled operations include hours that these should not be operating. We can now schedule them at specific times. Control of these kinds of devices is either a switch closure or a motor to turn on or off. If it is a switch closure, either a direct connection to a TowerLCC or a connection using the Solid-state relay will control that device.

# Periodic Event Schedules

Another shortfall of the EFX-16 or TowerLCC is the inability to schedule periodic events. Periodic events are defined to be those things that happen at intervals. In this case, I mean fixed intervals, like every 30 minutes or once every hour. A good example would be something like a cab;e car ride which is scheduled to operate every half hour between 9 AM and 8 PM. This would require 23 discrete events for the EFX-16. Another example would be a clock tower that chimed the hours throughout the day. This could also require the chime to be varied dependent on the hour.

The basic data for these use cases is already available in the Fast Clock. **Time Events** occurring at the selected **Rate** are the source of a coordinated

Time. The basic time-keeping function has also been demonstrated in the Lurker described in the previous chapter. All that remains is testing the current time and if it meets the interval criteria, do the required action. Some details are provided in Appendix B, including links to Github repositories for sample code.

## Summary

If the only things changing on your layout is trains, that is okay, but it is more exciting to see lights come on and off and other things moving. With the Fast Clock these changes can be synchronized with the time of day for realism. Using the methods and devices in this chapter allows these actions to be taking place while the operator is running a train. The EFX-16 is a great tool for lighting control and effects. Add-on drivers and solid-state relays are shown to expand the capabilities of the EFX-16.

# Chapter 11 — Trains, Throttles, LCC

Recent activity in LCC has been to extend the use to Train Control. From the first, there have been "Well-known Events: that were intended to be used for train control messages. Now some of these are being implemented. Train Control uses the concepts originally associated with LCC to send messages to train.

In LCC, messages are addressed to Nodes. Node addresses are unique to ensure messages are received by the intended node. In Train Control, this is used to implement a peer-to-peer network where one node communicates with just one other node.

Nodes in LCC are the basic entity for implementing. Every node has a unique address, some type of processing and data associated with it. Nodes can be either physical or logical. A physical node is one that is attached directly to the network through some network interface like CAN or WiFi. Logical nodes are contained within a physical node. Logical nodes are normally internal processing that connects to the LCC network through some other program in a physical node. Physical nodes are limited in number by the characteristics of the node, such as number of nodes for CAN. Logical nodes are only limited by the processing speed of the host. For clarity, unless otherwise stated, all nodes in the following are logical nodes.

In LCC, every train is a node, with a unique node address. Train nodes can be either real physical nodes or virtual nodes. Real physical nodes are implemented in the actual locomotive. This is probably a WiFi connected locomotive which runs independent of the DCC signal on a track. Physical Train Nodes have to perform all the LCC network functions independent of any other processing. These nodes are more likely in larger scales where the locomotive is battery powered and has sufficient room for the interfaces and processing required.

A virtual Train Node is currently more common. This is a node contained in a Command Station which converts the Train Control messages into DCC messages to send on the track. The command station needs to implement the additional information required. Connection to a throttle and matching a throttle to a train is also contained in the Command Station. A consist can also be a virtual Train Node, because the consist is the entity that is controlled by the throttle. Train Nodes are defined to be anything that is connected to the LCC network that can be controlled by throttle nodes. In this definition, passenger cars with lighting controlled by a decoder can be considered a Train Node.

A throttle is another node in LCC. Throttles connect to the Train Nodes in a peer-to-peer network. Each throttle is connected to a Train Node to control. In LCC, this connection is a two-way path. Train Control messages are created by the throttle and sent to the Train Node. In the case of a virtual Train node, the Command Station translates the LCC messages into DCC messages to send on the track. Throttles are connected through an LCC network, either wired or RF. LCC defines the messages for Train Control. Any throttle that sends the LCC Train Control messages on an LCC network will serve as a throttle. This standard allows multiple manufacturers to build throttles and command stations for Train Control.

## Overview of NMRA Standards

Three NMRA Standards cover the LCC protocols for operating trains. These protocols are inter-dependent and also build on the protocols currently in use. The operation of an LCC controlled train uses both the Event Transport protocol and unique LCC messages for Train Control. The three protocols are Train Control, Function Description Information and Train Search. Details of these protocols are documented in the NMRA Standards and Technical Notes. These details are not necessary for using the Train Operations.

### Train Control Protocol

To define train operations this standard first defines the terms needed. Terms such as Train and Throttle are included with the definition of a Controller and a Listener. Only one Controller is allowed for each Train Node. A Listener is defined as a Throttle that mirrors the Controller to watch in a supervised situation. The ability to "steal" the Train Node is also defined. A Train Node responds only to control messages from the Controller, but sends all responses to the Listener.

The messages necessary to provide train control are defined in this standard. The first section lists the events that support train operations. These include Emergency Stop, Clear Emergency Stop, Emergency Off All, Clear Emergency Off All and Is Train. The last one, Is Train, is used to identify all the Train Nodes on the network. During power-up, a Train Node sends a Producer Identified message with the IsTrain EventID. When needed, an Identify Producers request sends a Producer Identified reply with the EventID IsTrain. In this way, a Throttle can find all the Train Nodes currently available.

The specific action taken when the other EventIDs are received is up to the manufacturer. From the wording, the assumption would be to send the Emergency Stop to all Train Nodes to stop all movement. Emergency Off would be to shut off power to the track, but not all manufacturers are required to implement this.

Train Operation is implemented using the Train Control Command message for Speed/Direction, Set Function and assigning throttles as Controller or Listener. Since the LCC network is both Producers and Consumers, it is also possible to query these settings.

### Function Description Information

The protocol for Function Description Information supports Train Operation. The intent is to provide a definition of the action when a Function key is activated. This is then displayed on the throttle. The information is stored in the Train Node in XML format. A Throttle uses the Memory Configuration Protocol to receive the XML data for a specific Train Node.

### Train Search Protocol

Identifying a specific train on the LCC network uses the Train Search Protocol. This protocol uses already existing protocols to request all the IsTrain Nodes and then selects the nodes that match a 1 to 6 digit number in their descriptions. The request for a match is processed by the Train Nodes on the network. Each node compares the request with its internal information and returns a reply if a match is found. The requesting throttle then can request a search for a more complete description or select one of the matching nodes. If no reply is received, after 200 milliseconds, a Command Station may allocate a new Train Node to match the request, and send a producer identified message so the throttle can take control.

## LCC Command Stations

In the description of the Standards above, the concept of a virtual Train Node is used for those decoders that are not fully LCC. In fact, most existing systems like DCC do not provide the data or processing required in the decoder. In DCC, typically a Command Station is used to connect a Throttle to the track and the command station generates the DCC waveform required to operate a train. In this form of LCC, Command Stations are also required to translate the Train Control Command messages to the DCC waveform. In addition, the LCC Control Station becomes the host for the Virtual Train Nodes. The data required for multiple Train Nodes is kept in the Command Station and the Command Station provides the processing to respond to LCC messages. Since several Train Nodes may exist simultaneously in the Command Station, a Train Roster is created to represent all known Train Nodes. Since this is in the Command Station it is available to any LCC Throttle that requests a Train to control.

In LCC, the Command Station also supports all of the LCC Train data and interaction. The Virtual Train Nodes in the Command Station must respond to requests for identification, Function Key information and current configuration. Since a Train Node may be a consist, the Command Station also must implement consisting. An interesting part of this is that the Command Station data contains the decoder address. The decoder address must be unique and is not the only way to identify a locomotive. The identification information in the node is the primary method to specify which locomotive to operate. Using the Train Search Protocol, any field that has the requested search pattern will be provided as a potential train to operate.

### How Does It All Work?

When a operator picks up an LCC Throttle and decides to operate a train, what do they do? First, assuming the train is on the track, the operator starts the Throttle, connecting it to the Command Station. (I am assuming the locomotive is an older one with a standard DCC decoder.) The operator will enter a number to initiate the Train Search. When the Throttle receives the list of trains available, the operator will choose one from the list and send a request to operate that train. The train configuration is set to that throttle and the operator can control the train by adjusting the speed and direction. To learn what Function Keys are available, the operator may request the Function Description Information which will be displayed on the throttle screen. After a satisfying period of operating the train, the operator will stop the train (preferably in a yard or siding) and the throttle can be turned off to save the battery.

The preceding is the normal flow of control using LCC. A few optional cases are supported. First, if the locomotive does not exist in the Command Station, the operator can enter a decoder address. If the decoder address does not exist, the Command Station creates a new Train Node, using default information provided. After creating the new node, the operat1or may have to provide additional information to fill out the Function Description Information and a more identifiable description. If not, the train will be configured to run, just like the case above.

## Current Implementation

At this time (Early 2025) the only full implementation currently available is the CS-105 Command Station and the UWT-100 Throttle sold by Train Control Systems. An App for the Apple® iOS system on iPhones and iPads implements the throttle portion. I will use these devices in the following sections to demonstrate the functions. These devices may be partial implementations at this time, but the capabilities are mainly available.

## Exploring the CS-105 Command Station

The first thing to note when ordering the CS-105 is that it comes with a choice of power voltage and that the CS-105 provides power to the LCC network. If you have multiple nodes with additional power injection, it would probably be best to get the 15 volt power. This would make it compatible with the RR-Cirkits nodes. If you are using SPROG nodes, the lower voltage – 12 volts – is fine. Either one will work with LCC.

Hooking up the CS-105 is super easy. Plug in the LCC network cables, the track is connected to two terminals and the power is plunged into the box. The Quick Start Guide shows all these connections. Note that the Track Power comes in two types, Operational and Programming. For the next few things you do not need to hook up to either.

Let's look at the configuration options for the Command Station. The Command Station shows up as a Node on the LCC network. Open it just like any other node and you will see the Configuration Data as shown in Figure 11.1. Several interesting options are shown. The first two selections are the standard identification entries. The first unique options are in the segment named System

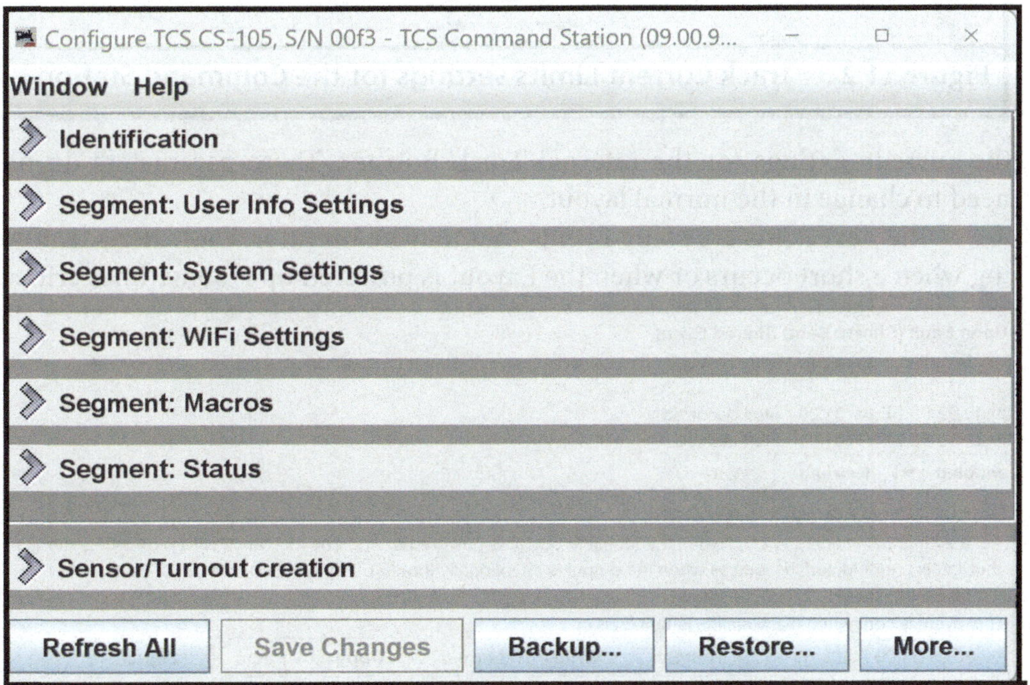


**Figure 11.1 CS-105 Configuration Options in LCCPro**

Settings. This segment controls some of the settings for track output, Fast Clock Settings and some advanced settings that are not common. These will be explored in the following sections. At the bottom are the familiar buttons for Refresh, Save, Backup, Restore and More. In the More button, there is a Factory

Reset option. Be aware that a Factory Reset will clear out all the Virtual Train Nodes, so you will have to re-enter that data.

The first section of the System Settings segment, shown in Figure 11.2, sets

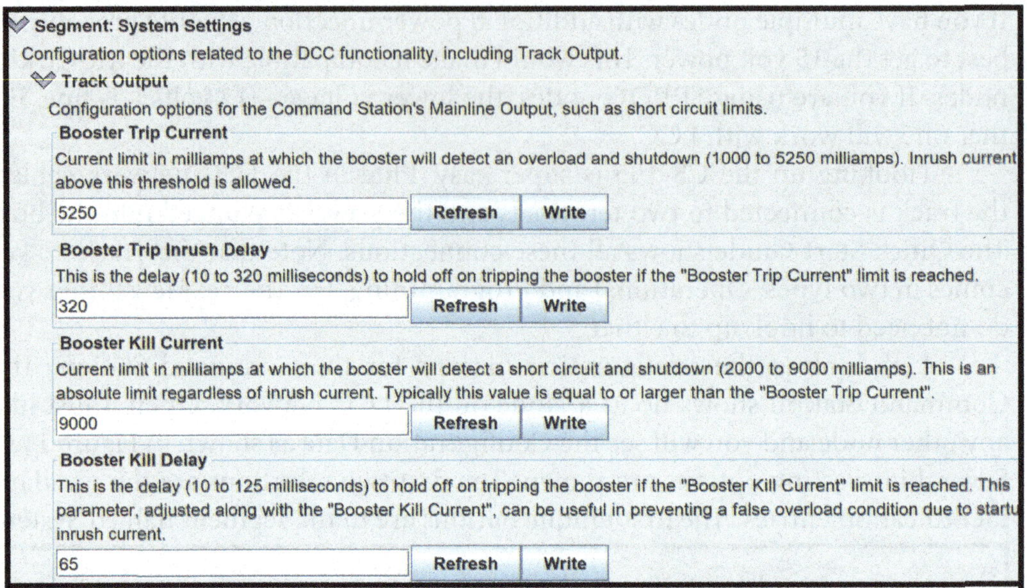


**Figure 11.2 Track Current Limits settings for the Command Station**

the current settings for the internal Track Booster. These parameters do not need to change in the normal layout.

The next items are some Events that may be useful in Fault alerts. Knowing when a short occurs or when the Layout is powered up is sometimes critical

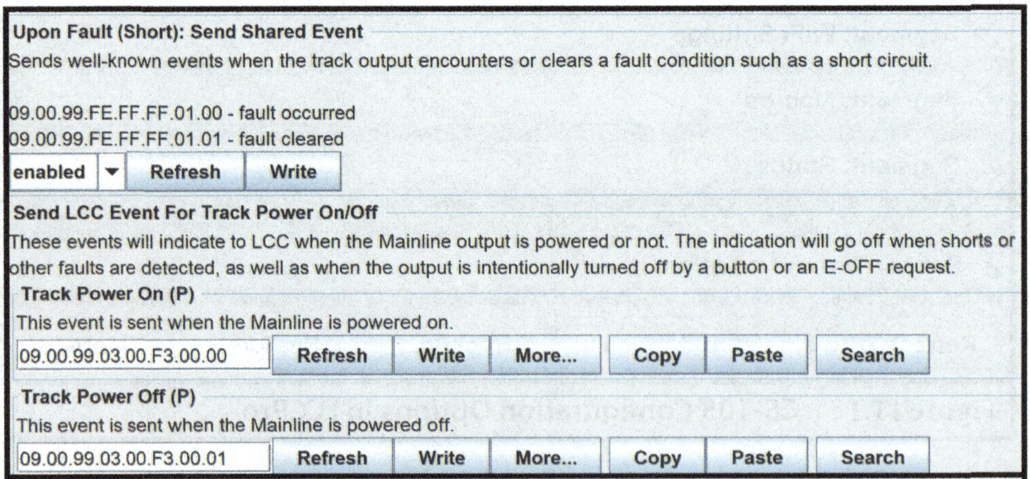


**Figure 11.3 Faults Cause some Events**

to the layout health.

The next segment is the Fast Clock implementation. It uses the same LCC Protocol as discussed in Chapter 9. In my experimenting and investigation, this is only a partial implementation and is difficult to control. For example, only startup parameters are included and there is no easy way to start or stop the clock. What it does that is useful is act as a Slave Clock to another clock on LCC. Then the time can be displayed on some throttles.

Advanced Options is the next segment in this area. The Advanced options cover such things as RailCom, Heartbeats, DCC Interface and Programing Track parameters. At the end of this section, there is an option to disable the Booster signal that is carried on the LCC network. This would then free up pins 4 and 5 for other uses. This is the only place that the Booster signal is mentioned, but now we know where it is.

The last segment in the System Settings is named Aux Cabs. This set of options allows use of CAB Bus devices. If your layout uses the CAB Bus, this may be of interest.

Moving on to the next major segment of the CS-105 Configuration, the details of connecting to the command station are shown. The WiFi segment shown in Figure 11.4 selects the mode of the command station and what networks it looks for. The first option is to set the mode of the command station WiFi to either Access Point or Client. As an Access Point, the CS-105 advertises itself to throttles as a small hub. It can support up to 4 simultaneous throttles through its WiFi module. In Client mode, the command station becomes a client on an existing WiFi network. Throttles then sign in to that network and

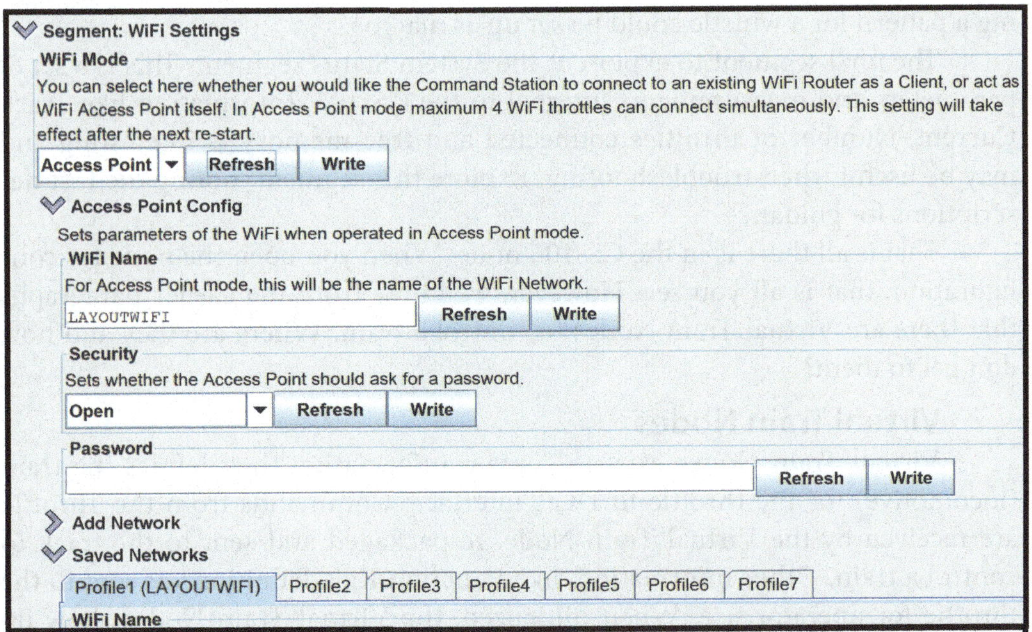


**Figure 11.4 WiFi Settings Control How Throttles Connectr to the CS-105**

these messages are routed to the CS-105. The limitation on throttles is now limited by the router, so it will generally be much higher. To use Client mode, refer to the CS-105 Quick Start Guide and the Throttle documentation.

There is another way to connect to the CS-105. If JMRI is operating and an LCC Hub is started. The LCC Hub will connect to the existing WiFi router on the computer. Messages received by the hub are sent on the LCC network and then acted on by the Virtual Train Node to control the train. To connect to this hub, the throttle must be signed in to the WiFi network, but the CS-105 can continue to act as an access point, giving the user two options for connection.

The name and parameters of the Access Point are the next optional fields. You may note that I have opted to leave my access point "Open" with no password. That is because I have nothing on this network that stays on or is sensitive. I am using this just for testing.

The next section in the WiFi segment defines networks to connect to in Client mode. Apparently the first one in Profile 1 cannot be changed. This may give an error message when exiting the configuration, but it can be ignored. There is space for 7 profiles, so using the command station on multiple layouts is supported.

The Uplink section contains parameters that are seldom necessary. Most of these will work without modification.

The next major segment is titled Macros and contains options to define Macros that can be initiated on startup or from a throttle. Macros are convenient for repetitive tasks. Tasks like setting several turnouts for a route or Sending a pattern for a whistle could be set up as macros.

The final segment to explore is the System Status segment. This is a set of parameters and measurements internal to the CS-105. Information like Track Current, Number of throttles connected and free memory is interesting and may be useful when troubleshooting. Explore this segment, noting the text descriptions for guidance.

That is all there is in the CS-105 node. When you open the node for configuration, that is all you see. However, we know from the earlier paragraphs that there are Virtual Train Nodes to control a train. Where are they and how do I get to them?

### Virtual Train Nodes

Virtual Train Nodes exist to contain information that defines the train (locomotive) for the throttle to DCC interface. Commands from the Throttle are received by the Virtual Train Node, re-packaged and sent to the track to control a train. Other information, like Function Key definitions, is sent to the throttle for operator use. When initialized, the Virtual Train Nodes show up on the Node lists in JMRI. There does not seem to be any way to create a VTN (Virtual Train Node) in JMRI. The only way to create a VTN is to select it in a Throttle by its DCC address.

Once the VTN is created in the Control Station, it will show up as a separate node on the LCC List of nodes. In Figure 11.5 three VTNs are listed, an SP with road number 4449, a UP GP40 with road number 501, and an unnamed Locomotive with decoder address 5402. The last one is the most recent addition.

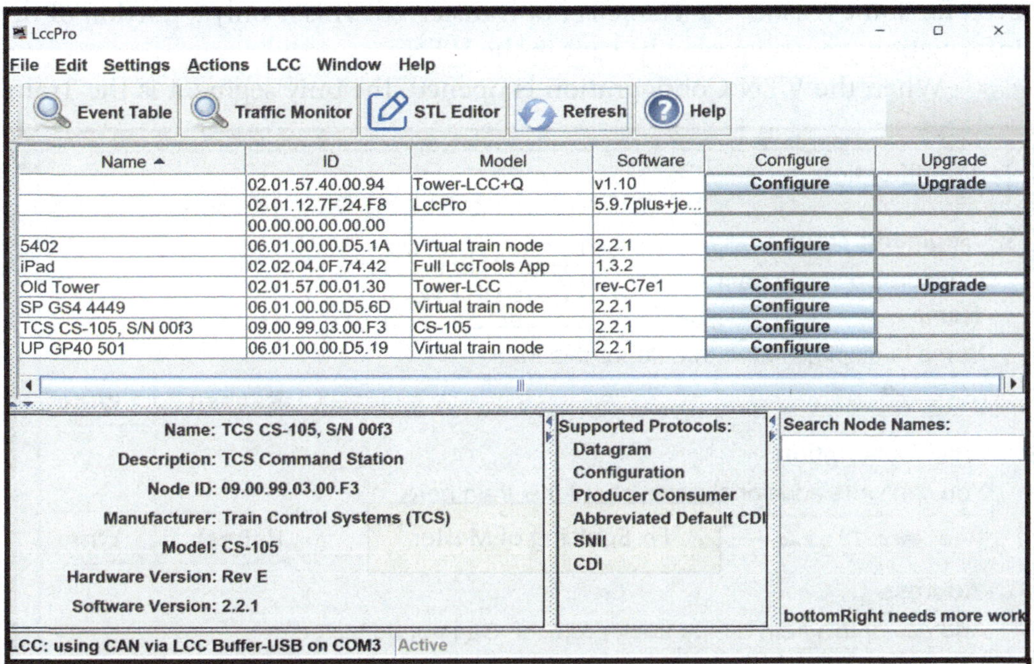


**Figure 11.5 Virtual Train Nodes (VTN) Appear on the Node List**

I can open the configuration data for this one and add information. Of course, I can also open any other VTN and change the information in that VTN.

A couple of things about the VTNs. Creating a VTN by selecting the DCC address seems to be the only easy way to create one. Because of the Train Search Protocol, if you enter the DCC address of an existing VTN, you will be linked to that VTN, not to a newly created VTN. This ensures that there is one and only one VTN with that DCC address.

The second point about entering just a DCC address is that they are apparently temporary. If I do not add a name for the VTN, powering down the CS-105 will erase that VTN. This is convenient for visitor use because the added VTNs do not stick around to clutter up the node lists. Node lists will become cluttered just with the locomotives one individual has. In a club, the number of locomotives may easily reach several hundred. Maybe this is the solution: after each day's operations, the Command Station is powered down to clear all the VTNs for the next session.

When creating a VTN, many times there is already some information in DecoderPro for that locomotive. DecoderPro has an Export feature to transfer some of the data to the VTN. To access this feature, open DecoderPro with the

Connection in Preferences set to the LCC Buffer. Then open the locomotive in the Programmer, pull down the Files tab, and go to Export. When it opens up its options, one of them should be Export to TCS CS-105. After selecting it, wait until the pop-up says it is finished, and some information will be filled in. However, for some reason, the Name is not transferred. This is only a portion of the information you will probably want in the VTN.

When the VTN Configuration is opened, the only segment is the Train.

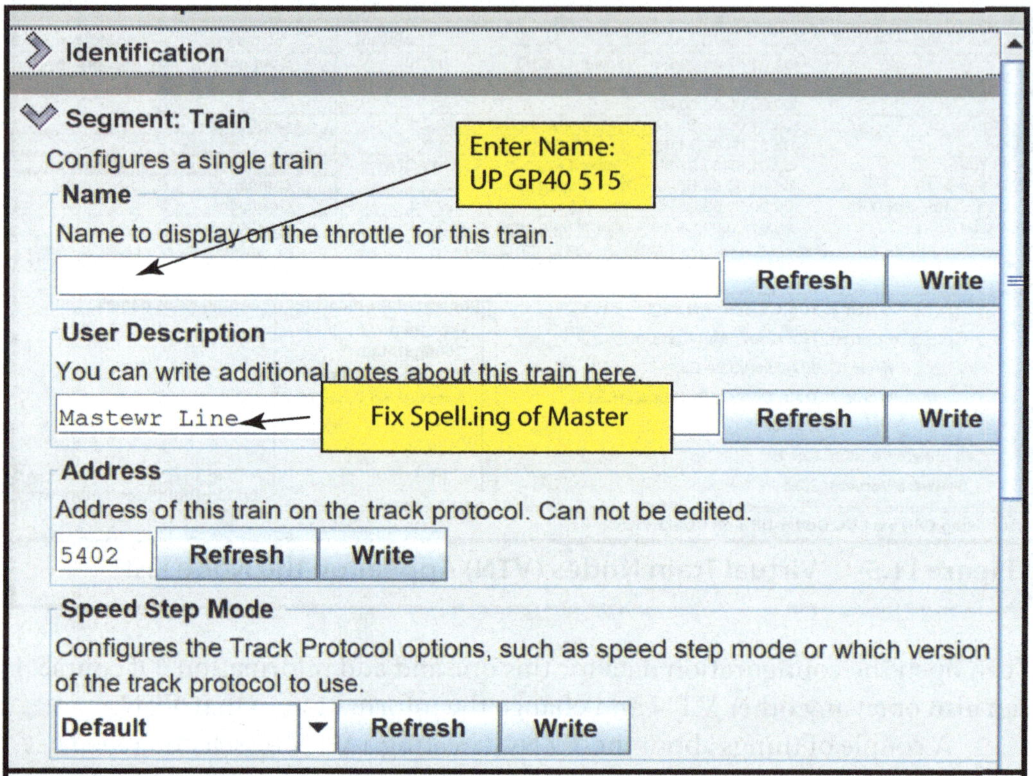


**Figure 11.6 The VTN Node Options and Data**

Expanding that segment displays the initial information to fill in. In Figure 11.6 there is some information already filled in because I imported it from DecoderPro. The rest of the VTN needs to be entered manually. The two fields called out in Figure 11.6 are the only fields that need to be filled in to save the VTN. Notice that the Address field cannot be modified.

The next segregant is concerned with the Headlight function, Function Key 0. Until you want to fine tune the actions of a headlight in a consist, the defaults in this should be fine. Notice that this function key is not modifiable. All decoders have the Headlight connected to F0.

The most work in defining a VTN is in the Function Key Mapping. This feature defines the names for the Function Keys. These Names are then displayed when requested on the throttle. Throttles are often limited in the number of keys that are available, but there are ways to stretch to activate these functions. As shown in Figure 11.7 there can be up to 28 function keys defined in a VTN. In the example shown, the F5 key is being defined as the Rotary Beacon.

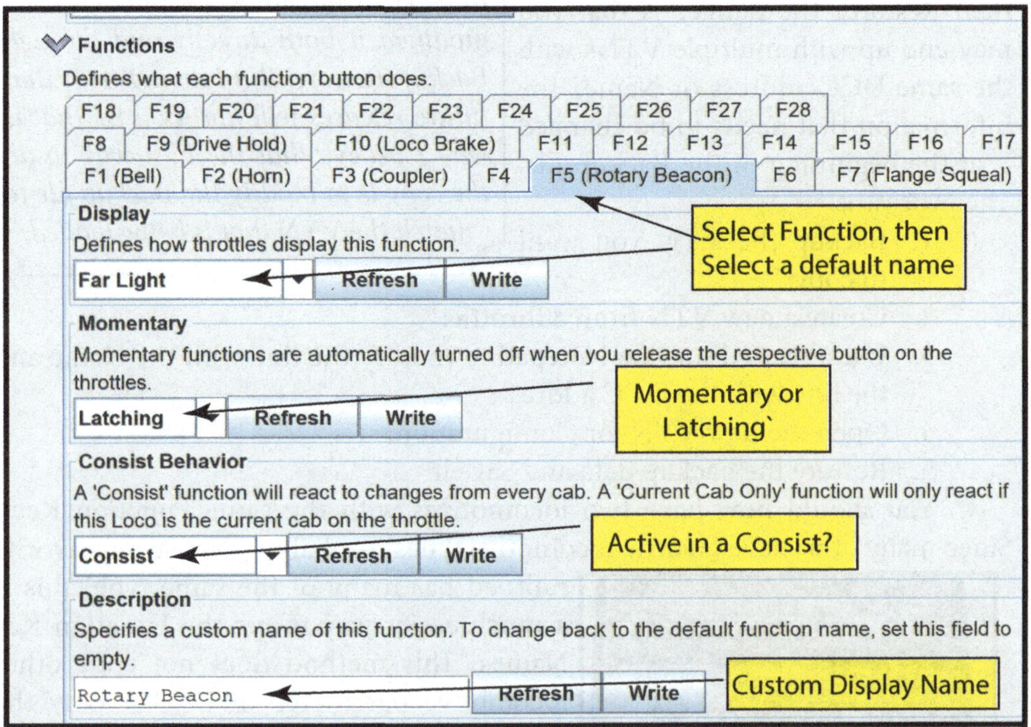


**Figure 11.7 Function Key Mapping so the Throttle can display Names**

Since that name is not one of the Defaults available, the default chosen is "Far Light." At the bottom of this screen, the Description field is entered as "Rotary Beacon" so that the display name is correct. The tabs for the Function Keys display the names entered. In this case, several of the Function Keys are using Default names. If the Default names do not appear in the tabs, simply "Refresh All" and they will be filled in. A detail that may need to be fixed is the Consist Behavior. If this locomotive is used in a consist, it would probably not have a rotary beacon on every engine. The Consist should have a "Cab Only" option selected.

The last item on the VTN configuration is the Delete From Roster option. If you want to remove this VTN from the CS-105, select "Delete all Settings" and close the Configuration window. After a power cycle, the VTN will no longer exist. All the work will be lost. Of course, there is also a Backup and Restore, so you could have backed it up. To restore it, you will have to create a replacement with the DCC address, then open its configuration to restore.

A slightly dangerous method to add identical locomotives (with identical decoders) without the effort of setting all the Function Key Mappings is to Backup the VTN that is complete, and then modify the Backup file to the new locomotive name and DCC address, then Restore. The danger is that you may end up with multiple VTNs with the same DCC address or Name. The information that needs to be changed is at the beginning of the Backup file. The process is:

**DCC Address Constraint**

*At this time, the DCC Address parameter claims to be read only. I have modified it both directly and through Backup/Restore. But I do not trust that future releases will not fix this. That is why I believe that the bast way to do the copy is to modify the Backup file to match the VTN that is being loaded.*

1. Backup the VTN you want to copy;
2. Create a new VTN from a throttle;
3. Use a text editor like Notepad to modify the line with the Name and the line with the DCC address;
4. Open the new VTN for Configuration;
5. Restore the backup data and Save it.

You should now have two locomotives with the same Function Keys. Since many of us have favorite locomotive types (probably because our favorite railroad has many of the same type) this is a much easier way to get the Function Key Names. This method does not copy other locomotive parameters like CVs, only the Name, Address and Function Keys are copied.

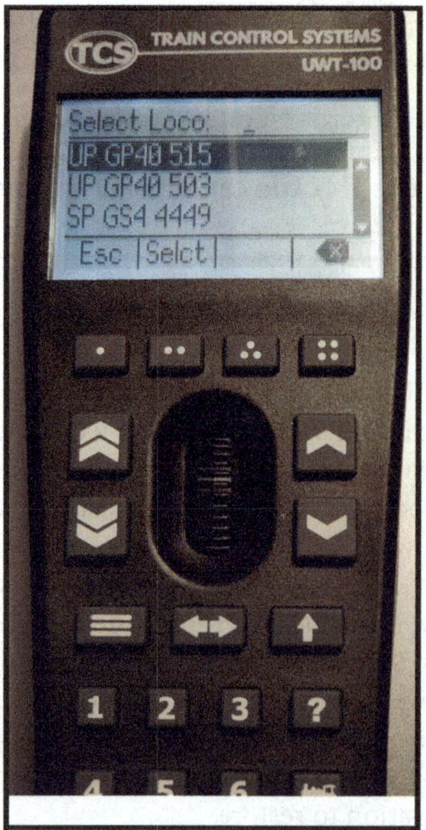


**Figure 11.8 The UWT-100 Displays Trains to Select**

The CS-105 is the Command Station for the DCC on the track. It also supports RailCom and CAB Bus. I have not investigated those to keep this simple. If you have the need for them, you will have to explore that on your own. This will get you started in the use of LCC. TCS also has as Booster available to increase the power for multiple locomotives. The Command Station uses the LCC Cable – pins 4 and 5 – to send the DCC signal to the Boosters. The Boosters also are connected to the LCC network and have events like the Command Station.

## LCC Throttles

Throttles that use LCC as the interface to a Command Station are not currently widely available. I expect that to change. The advantages of using LCC messages and interactions will attract users and manufacturers will respond. One manufacturer, MRC, has announced devices that will use LCC messages. At the time that I am writing this book, three implementations are available, the UWT-50 and UWT-100 by TCS and a iOS implementation which will run on Apple devices. I will describe the UWT-100 and the LCCTools implementations, primarily describing the LCC functions.

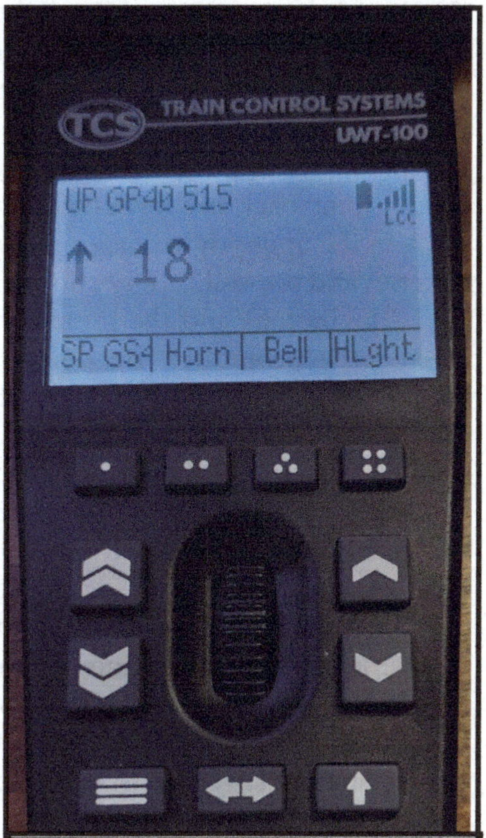


**Figure 11.9 Operating the Train with the UWT-100**

**UWT-100** is a traditional type throttle on the outside. It is a plastic case with a numerical keypad augmented by additional keys, a thumb-wheel and a small screen. It operates on two AA batteries and even has a flashlight mode for night operations. It is strictly wireless and connects to an access point or WiFi router. Connecting it initially is covered in the Quick Start Guide. You may want to modify the CS-105 to use a local WiFi. As initially configured, both Command Station and Throttle assume use of the CS-105 as an Access Point. If the connection is not acquired, the throttle will ask to Add a network. If this happens, a list of available networks will be displayed. Using the Up/Down arrow keys and the button under the screen labeled Select, a network can be selected.

Once the UWT-100 is connected the operator needs to select a locomotive. On the screen shown in Figure 11.8, the available VTNs in the CS-105 are listed on the screen. Using the Up/Down arrows, the desired locomotive can be highlighted and then selected by pushing the button labeled Select. If the locomotive is not listed, it can be created by highlighting the line on the upper Right and entering the DCC decoder number. When the ADDR button label appears, press the button. As noted earlier, this causes a new VTN to be created. It will be only available until the CS-105 is turned off, so this is great for guests. Pressing the ADDR button ensures that the decoder address will be used for the search, not any other number found.

A frequent action is the creation of a consist. This is quite easy, just press the Menu button (three horizontal lines) and step down to the Consist section. Selecting this will allow you to Add to the current locomotive or consist, list the consist or clear the consist. In the List, you can clear a selected locomotive. In addition, functions can be restricted to just the leading locomotive.

Now that the locomotive (or consist)is selected, the train is ready to run. The throttle screen shows the current state of the train. Most of the features and controls are clearly identified. The screen in Figure 11.9 shows train "UP GP-40 515" moving Forward at a speed of 18. Across the bottom of the screen there are three function which can be triggered by pressing one of the buttons below the screen. The first button is labeled with a locomotive name. This allows you to quickly jump back and forth between two trains or consists.

In the upper right corner there are some icons that show the status of the batteries and the WiFi connection.

The buttons and the thumb-wheel in the center are used to control the speed of the locomotive. The double arrows on the left jump the speed 10 steps at a time, the single arrows on the right step one step at a time. The thumb-wheel in the middle moves one step per click.

Below the speed controls there are three buttons for the menu, direction and a shift button. As stated earlier, the menu button will list additional features, like consists, Operation information and help. An important option in the menu list under Operations is the Release option. This disconnects the throttle from the train. Explore other parts of the menu to find what you will use.

The first three Function Keys are on the screen with names. All the other Function Key Names can be listed under Menu->Operations->LocoFunction-List. These are the names added in the VTN in the CS-105. When operating, use the number keys to activate these functions. If the number is greater than 9, press the Shift Key then the number minus 10. For 20 to 28, press the shift key twice. A small number 1 or 2 will appear to tell you are in that range.

These are the most important features from a LCC point of view. The throttle has the ability to retrieve and display locomotive names using the Train Search Protocol. It retrieves and displays Function Key names. It connects to the command station to control trains through VTNs. The rest of its capabilities are mostly related to running trains. These capabilities are common to most command stations and Throttles.

**LccTools** is an App for Apple iOS devices. It originally was written to provide a Configuration Tool on an iPhone or iPad. The LCC Throttle portion of the app provides another throttle to interface with the CS-105. LccTools is a wireless device, so it needs a WiFi connection to operate. For the CS-105 this can be the internal Access Point, or JMRI can start an Lcc Hub to transfer LCC messages from WiFi to the CAN network. Top start the Hub, click on the LCC tab at the top of the LCC window. An option for starting a hub should be on the pull-down list. Simply click it and an LCC hub will start.

Before you start the LccTools App, check that the WiFi Setting has the correct network open. Either the CS-105 Access Point or the WiFi Router has to be connected to the computer running JMRI. The LccTools Settings screen in Figure 11.10 verifies that a connection is present.

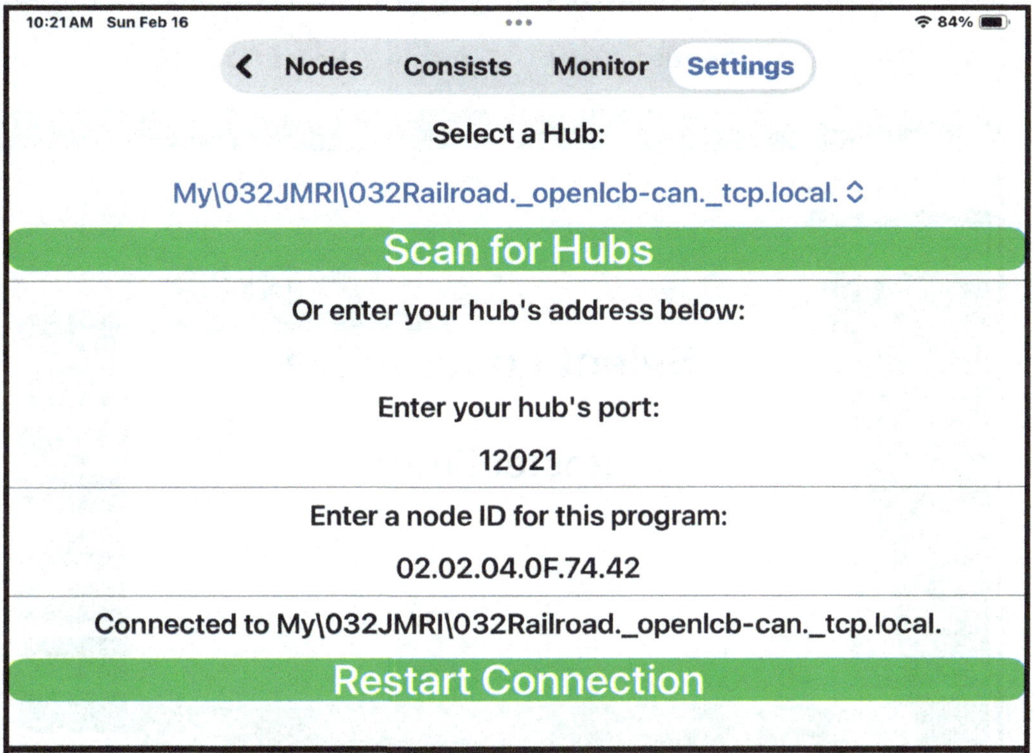


**Figure 11.10** LccTools Indicates that the Connections are Setup

When LccTools is ready, the tabs across the top provide access to various features. The set of features on this screen include a Monitor which is like the JMRI Traffic Monitor. The next tab is for creating consists. Like the UWT-100, a consist is created by entering a set of locomotives. Options on this screen include adding to a consist and removing .

At the far Left on this screen is a tab labeled Nodes. Nodes opens a screen with a book icon at the top. Tapping this displays another screen with a Remote Nodes in the upper Left. Tapping this label opens up a list of all the Nodes currently on the network. Tapping any one of these Nodes will display a screen with the initial data from the Node. At the bottom is a set of options for displaying data from the Node. Included is Configuration which will open all the Configuration data. This may take a little time.

Tapping the arrow at the upper left brings you to the Throttle Setting screen. Tapping the Select bar on that screen will open the Select screen where you can Select a specific locomotive to control. As with the other throttles, you have a choice between picking from a list or providing the DCC Address.

After you select a locomotive, the throttle control is provided by the screen shown ion Figure 11.12. Notice along the Right side, a list of Function Keys tells you what each one does. This is a convenience for the operator since the functions are listed in names not numbers.

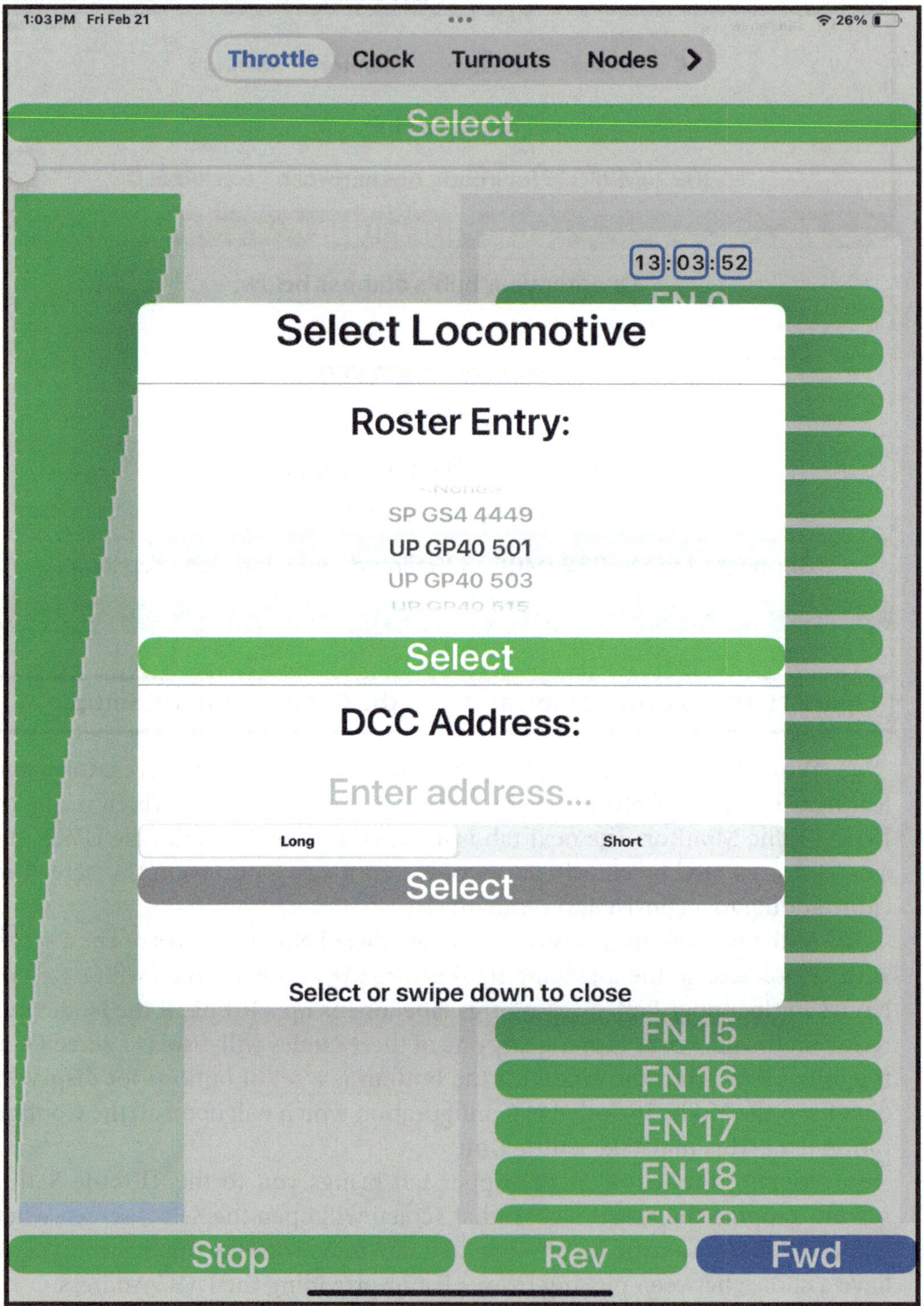


**Figure 11.11** Selection of Locomotives is either from a List or by the DCC Address.

Controlling the speed is either by tapping the left side to move up or by using the slider at the top of the display. Foreword, reverse and stop are at the

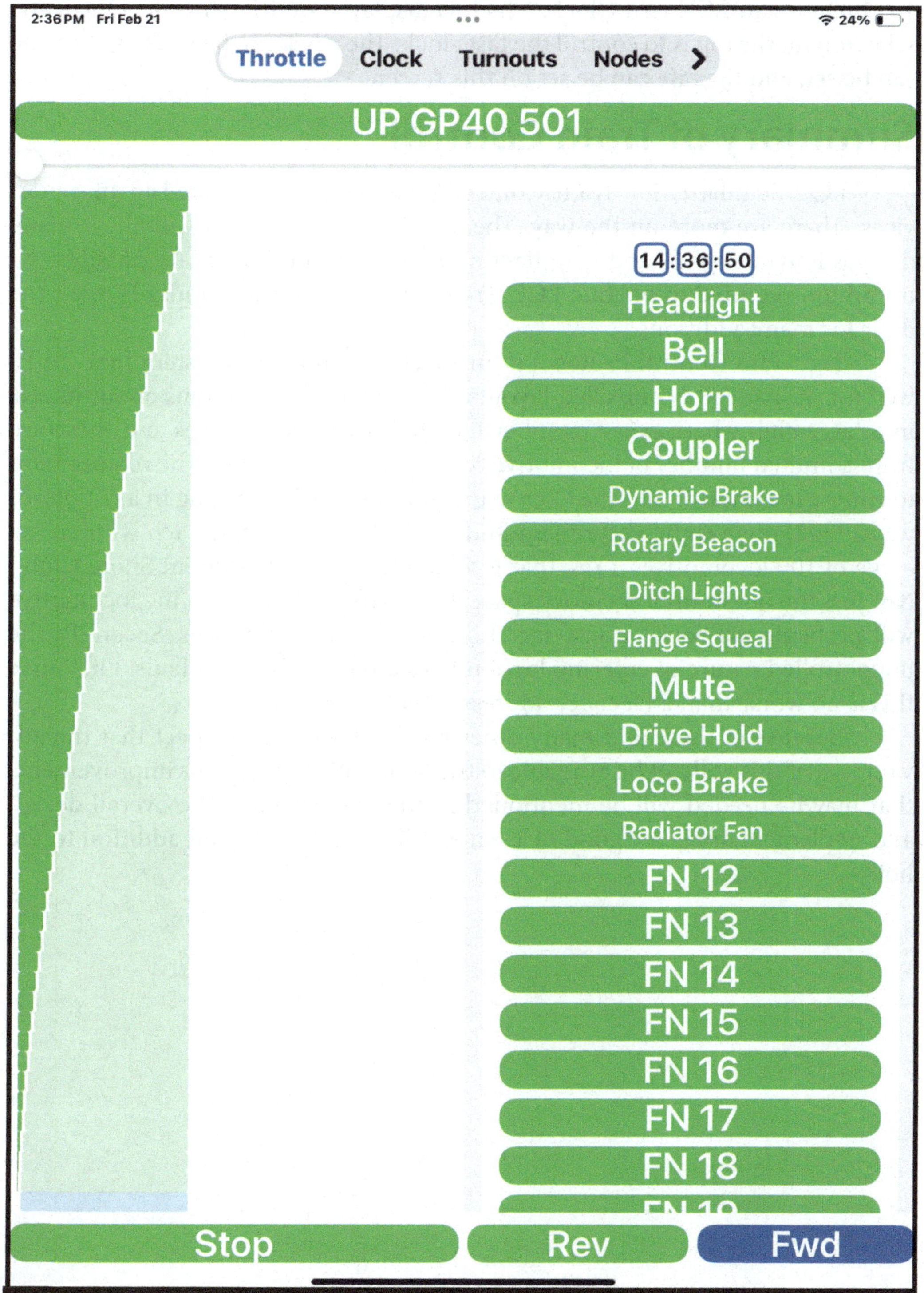


Figure 11.12 The Throttle Controls a Locomotive or Consist

bottom. As the speed is increased, the left hand side indication becomes Blue, giving a quick indication of the speed.

Notice at the top of the Functions there is a small clock display. The CS-105 includes an LCC Fast Clock, which is displayed on this throttle. One if the selections at the top is to control the fast clock. The clock can be paused, the time can be set, and the rate can be set on this screen.

# Summary of Train Control

LCC Standards for Train Control have been implemented in these devices. There are more on the way. The creation of a Standard will allow many devices to be designed and manufactured. The current devices are possibly the minimum required to operate LCC Train Control, but the Standards open the door for many additions.

The Train Control Protocol is an excellent guide for a system that works well for individually controlled layouts. I have some reservations about its use in a large club where every member has their own locomotives and these are from a limited number of locomotives commercially available. The requirement to enter a locomotive into the Command Station may be limiting in a couple of ways. First, there is the sheer magnitude of 30 to 50 members each with an average of 100 locomotives. (Yes, that is what we see at the Belmont Shore Club.) Not just the question of memory space, but trying to find a specific locomotive will probably fall back on just searching for decoder addresses. Secondly, the uncontrolled nature of entering locomotives could lead to problems. Of course this is no worse than a Decoder address oriented approach.

Because of these and many other required actions, I expect that the advantages of this will not be available to the large club. Some other improvements that may be needed will be mentioned in the next chapter. The overall design and implementations are good or even excellent. This is a good addition to the hobby.

# Chapter 12 Present and Future Options

LCC is a vast playground for Model Railroaders. This book has concentrated on LCC as the primary method for implementing control of model railroads. Several alternatives have been bypassed to keep the dialog clear. Some of these are noted in this chapter. In addition, this chapter identifies some future possibilities that I think will pop up. I do not have an accurate crystal ball so I suspect many other things will show up.

## Present Options for LCC

Options for the implementation of LCC are primarily add-on or other ways to perform the tasks discussed in the previous chapters. Many of these are personal preference. Trade-offs between cost, performance and time to implement are generally how the decisions are made.

### JMRI Support

Before LCC, there was JMRI. In many cases, JMRI has provided the control for model railroad layouts. In this book, the only use has been to configure nodes. Many modelers have used JMRI for signaling and throttle connections.

In LCC there are two places where JMRI can be used besides the Configuration Tool. The first that has been somewhat integrated into LCC is the creation of the sensor and turnout tables used by PanelPro in building a control panel. Most configuration windows have buttons to generate these tables from the data in the node. My club is not currently using PanelPro because there are too many operators working independently. Future plans include a Dispatcher station for CTC which will probably be implemented with PanelPro.

Another support function that has been implemented in JMRI is a STL Editor. Creating the STL expressions in JMRI lets the user reference the Input/Output, memory and Track Circuit variables using names. Then when the information is uploaded to the TowerLCC, the names are replaced with the table references used internally. There does not seem to be any syntax checking, so the syntax checking is in the TowerLCC. The syntax is more likely to be correct, but if it is not, finding the error could be difficult. The STL editor in JMRI has some of the same problems that are in the User Manual, specifically, the FR operator is still in the operator list. I do not see a great advantage, because you still have all the restrictions. In fact, since the STL editor is a table with one operator per line, I find it difficult to read. I prefer a full conditional expression to be together.

### RR-CirKits Loconet Gateway

For those layouts that have existing Loconet® Accessories RR-CirKits has implemented a gateway. The idea is to identify EventIDs that are intended as accessory commands and exist on the Loconet network. Then the gateway will convert the EventIDs to Loconet commands. The user has to connect EventIDs to Loconet commands in the Configuration of the gateway. This device is a step in converting from Digitrax® to LCC. The gateway operates in both directions, so you can also use Loconet sensors for LCC information.

### RR-Cirkits Tortoise Quik-Link

This is the first of several "Quik-Link" devices. Quik-Link is a three wire connection for easy connection to simpler devices. RR-CirKits sells a breakout board that provides a set of Quik-Link connectors for each of the lines in a 10-pin IDC connector. This allows a user to connect a Quik-Link device to a node with a servo style cable. The Tortoise Quik-Link is a driver for the Tortoise and does not require a SMD8 or other device to control a turnout. Connecting directly (through the breakout board) the device is a stand-alone turnout controller. It is more expensive per turnout that an SMD8, but would be useful in some cases.

### RR-CirKits Button Quik-Link

Another device using Quik-link is the Button Quik-Link. This is a self contained illuminated push-button which operates in the sampling mode. It is advertised to be a simple way to make a control panel because there is no soldering required. Just connect it up to the Quik-Kink breakout board and you have all the illuminated button capability. There are a couple of issues, however. The first is that the ones I have seen come with a solder bridge (an intentional option) which has the button displaying an alternating red-green. Looking at the device in Figure 12.1, there are three silver blobs that are the solder bridges for

**Figure 12.1 The Button Quik-Link options are Solder Bridges.**

options. On the right end, the option is whether to alternate or not. As shown, this one will alternate between phase 1 and phase 2. Removing this bridge will cause the button to display phase 1 when pressed and phase 2 when released.

On the left end, the colors associated with each phase can be selected. As shown, the phase 1 color will be Red and the phase 2 color will be Green. Removing these bridges will cause that phase to be unlit. In many cases these options may be acceptable, but other options may be preferred.

Another issue with these buttons is that they are somewhat expensive. The features provided make them attractive for some cases, but for a simple control panel like the ones shown earlier, the cost per button is a concern. Balance that against the ease of use (even considering the solder bridges that may need to be removed) and you may find these buttons a quick way to make a control panel.

### Berrett Hill Touch Toggle

The Barrett Hill Touch Toddles are another button option that is compatible with the sampling mode of the RR-CirKits LCC nodes. These buttons are a touch technology, so there is no mechanical action. The buttons operate like a touch screen. They need to be mounted according to the directions. If you are a fan of touch screens, this is an interesting device. They can be hooked in to the RR-CirKits Quik-Link breakout board for easy connection to LCC. They are slightly expensive, but would be convenient on a fascia.

### Servo Motor Turnout Machines

Some model railroads use a servo motor mounting to control turnouts. The servo motor is mounted in a base that holds the servo motor and two micro switches. The micro switches are used for power routing to the turnout frog, and are mounted so that one side is activated when the arm is on that side. The turnout is controlled similar to a Tortoise with a stiff wire connected to the servo arm.

Servo motors are not stall motors. They have a limited angle of motion to travel. In use, the servo motor requires a special activating signal which is basically a PWM signal. Depending on the signal, the servo motor moves to a specific position. In current applications, many users have implemented these turnout controls using small Arduino boards. With the availability of the Sprog boards described in Chapter 3, servo motors can be controlled directly from an LCC node.

**Figure 12.2 A Servo Motor Controller for Turnouts**

The example shown in Figure 12.2 is mounted under the layout base. In the figure, it is upside down, with the wire going up through the base to the turnout. Some flexibility in motion is provided by the lever arms on the micro switches, but the travel should be adjusted so that the servo motor is not stalled and drawing too much current. Conversely, there has to be sufficient travel to force the turnout points tightly against the stock rails. Selection of the servo motor must consider the stiffness that must be overcome to reliably close the points. Servo motors are not all the same. Some may not be very strong, so making a full travel to press the turnout rails may be a concern.

Servo motor control has been promoted as cheaper than Tortoise machines. If all the issues are considered. I believe the result is that either way is about the same cost.

### Snowball Creek DIY

A recent announcement described a DIY node possibility using an Arduino μC. The device available for sale is a shield to mount the UNO μC which provides the interface to the CAN network. A library exists on the Arduino IDE to support this combination. I have not tried this, but it is a start into the area of DIY LCC on an Arduino. Other devices are in work, but this is the first commercially available.

## The Future is In Sight

### The Playing Field

Previous chapters have introduced the terms and processing of LCC. Since LCC is still developing new devices and expanding the functions through new protocols, we need to reflect back on what we have learned and how that may apply to future capabilities. Looking to future developments, the LCC standards have been designed for enhancements. Examples of places where the network can be expanded are:

1. The network protocol (CAN or TCP/IP) is not inherent in the design. Using a different network protocol only requires adapting the messages to the constraints (if any) of the network.
2. Event Identifiers currently occupy a small subset of the available identifier range. As additional capabilities are defined, additional Event IDs can be assigned without resorting to changes in the basic message structure.
3. Unique Event Identifiers have been given a large range. Running out of these identifiers is highly improbable.
4. Processing requirements are modest, given current technology. In the future,, faster and larger microprocessors may be able to expand the capabilities.

These design considerations ensure that much of what we have learned in the previous chapters will apply to future developments. In the development of nodes and configuration tools, the expectations for future products has also been established. Every new device should meet or exceed the capabilities in the RR-CirKits nodes and in the JMRI Configuration Tool.

The following list is what I believe is important in the previous chapters:

1. Event IDs, Node IDs are globally unique and are assigned by the manufacturer of the node. This guarantees that there will never be a conflict with a new board, regardless of manufacturer.
2. Networks will send and receive the same basic messages as defined by the current standards. This does not mean that the content of the messages will stay the same, but a node will recognize the structure of the message.
3. The Producer/Consumer model supports any conceivable interaction between the devices on the layout. The use of Events to control the layout allows a network that can handle a large layout with nominal message traffic.
4. Configuration Data is required to reside in the node. Configuration Tools like JMRI will read the data and display it in human readable forms, so that the node can be configured without physical access.
5. Nodes will have I/O similar to what we have worked with. The I/O types that were used in the previous chapters will be the same or similar in the future. Since the I/O is similar, the Node Tables used to define the I/O for configuration should be useful.
6. A distributed network prevents a single failure from disabling the entire network. Troubleshooting is enhanced because individual modes can be isolated to locate the failure.

In future devices, these items need to be examined to see if the device meets the criteria. Many of these are required by the standards, but others are considered good design practices.

### Going Forth into the Future

So far all of the devices that have been described are commercially available. Future developments are in work to expand the availability of LCC devices, and extend the functions of LCC. Three major directions seem to be currently active. First, the effort in extending the standards seems to be in the direction of defining new message protocols for specific capabilities, such as Location Reporting. Secondly, in hardware development, new devices for gateways and bridges are in development, as well as work to adapt the LCC to TCP/IP networks. The third area is a possibility of creating an environment (including software development environments and hardware) for Do-It-Yourself hobbyists. To keep track of all of these developments, the best place to go is the OpenLCB group.

*LCC Groups on Email*

*To stay informed and to get expert advise, two groups are established at groups.io. The first is the original OpenLCB group. This is where new things are discussed, and is more technical in scope. Changes to the standards are developed within this group The second, LCC, is more oriented to the installation and application of LCC. Both of these groups are hosted by groups.io. To join either of these groups go to groups.io.*

One current proposal is to define a Location Services Protocol. This is being considered as a Working Note. The idea would be to report that something (a train) passed through a location and a sensor (RFID, QR, ?) read the unique identification. An Event with that information is then sent. This proposal is still in the Working Notes stage where comments and improvements are collected. The specific implementation is not yet defined.

The following are things that I have thought would be good additions to the LCC collection of capabilities. They are certainly not the only things that could happen, just some that I think would be useful;. Some of these may be already in work and may be announced before this is finally published. Others may never happen. That is the way of wide-spread development.

### LCC Android App like LCCTools

Already implemented on the Apple iPhone and iPod, the App provides a wireless connection to an LCC network, including emulating a throttle for the CS-105. An implementation on the Android would increase the use of both LCC and the LCC Train Control.

## JMRI LCC Command Station

The advantage to creating an LCC Command Station within JMRI is that it could be a way to add LCC Features to an existing layout. If JMRI performed the Virtual Train Node Functions and then used the existing methods that provide DCC interfacing for hardware like Digitrax, NCE, Lenz, etc., a mixed layout could result. The point is that JMRI already talks to other manufacturers control buses so the DCC portion of an LCC Command station is already available. I would love to take my TCS throttle to the club and run my train on the Digitarx layout using the LCC Protocols.

## Power Management

A missing element in the TCS offerings is a Power Management device. Similar to the Digitrax unit, a power management device would protect the rest of the layout from short circuits. Extending the idea, the Power Management device would be connected to LCC for configuration – Current trip level, EventID when activated, etc. While these protections are currently available at the Booster level, many layouts like to sub-divide the power into shorter sections. The LCC Power Management/Short Circuit device would be configurable by LCC and would send an Event (configurable) when a short occurs.

## Block Management with RailCom

Extending the Power Management idea, if each block was provided with a small booster-like device that provided power to a block, that device could be a occupancy detector, a RailCom detector and a short circuit protector. I would recommend grouping several (4 or 8) in a single node so the number of physical nodes does not become too large. The cost would have to be modest, but the main work for DCC is at the command station. The DCC signal is created there, so this device is just a track driver, probably little more than a full H-bridge controlled by a μC. Since the DCC signal can be carried by the LCC network, it would already be available.

## RFID Reader

In the area of Location Reporting, an RFID reader is one of the possibilities. RFID Tags are ubiquitous these days, with many options. One of the options for the tag is a small, very thin sticker. It is small enough to put on the bottom of a locomotive. When the locomotive passes over a reader, all the information will be read into the reader. So what do we do with that information?

The easiest application would be to use the detection to lock or unlock sections of the layout. During an operating session or an Open House where visitors roam around, control panels may be locked to prevent mischievous button pushers. There are devices available to provide a on-off type signal when a RFID tag is waved over them. That signal can be used as an input to Lock and Unlock in this use.

More complex use would include the reading of a identification for tracking a locomotive. In this case, several readers would be located at critical points, and the reporting would include the location.

### Power Injection with Measurement

Every Node manufacturer seems to include a power injection capability. What is less available is the measurement of that power. I would like to see a Power Injection that included measurement and a trip point that can be set to send an event if the current is exceeded or the voltage drops. This is the simplest way to get a warning, but extending this to a true measurement, a current level or voltage could be sent. That would require an Event with a Payload, and few places would be able to display that.

Many more complex devices (like the EFX-16 and the CS-105( already provide power, and all power injection has the network in the connectors.

### A Bridge to connect Major Segments

Many layouts probably have significant localized sections that have heavy interactions. An example would be a major switching yard with LCC turnout control. Many of the Events will be just between nodes no more than 10 to 20 feet apart. This idea would be to have a nearly self-contained LCC network in this area and then use a Bridge to send the few Events that are not just local to the rest of the layout. The advantages would be less Event traffic on the rest of the layout and the ability to support more nodes with less distance.

## Do-It-Yourself (DIY)

DIY has always been popular in Model Railroading. Many hobbyists have built rolling stock, structures and track work. Bench work and scenery are almost always built by the modeler. Many also extend their talents to electronics, either building circuits or at least installing them.

Recently a DIY phenomena has revolved around the availability of a simple to use, easy to program micro-controller called Arduino. The Arduino micro-controllers define a specific set of I/O pins and provide a simplified programming environment with built-in functions for manipulating the I/O. Because these were originally developed for non-technical persons like artists, these programs are called sketches.

The Arduino system is popular because it provides that easy to use entry into a controller. It does this by some very clever things. First, the definition of the board pins allows multiple manufacturers to make add-on boards called shields. These shields provide specialized I/O capabilities like motor speed control, touch screen interfaces, servo interfaces, GPS receivers and many more. These shields come with libraries of code that allows easy use.

Second, the code for an Arduino is developed on an Integrated Development Environment (IDE) specific to the Arduino family of processors. Embedded in this environment are many sub-programs to make the interface to the hardware simple. Manipulation of registers and ports in the micro-processor are hidden, so the only thing the user has to program is the unique code that he/she needs for the specific job. Nearly all of the "boilerplate" of the micro-controller and programming language are provided without additional user programming.

The IDE also hides the functions of linking code together and loading it into the target micro-processor. This eliminates a lot of effort in learning how to do these steps and also eliminates the many errors when typing in complex commands. Basically, Arduino lets the user get down to the job, without worrying about the boiler-plate commands and steps. This should be the goal for DIY LCC.

## DIY LCC Modeled on the Arduino?

If a DIY LCC is advertised, what are the features that I would consider important. Using the Arduino as a guide, I would consider the following features as essential to the success of LCC DIY:

1. Availability of a capable micro-processor at a reasonable price. Currently, several manufacturers offer micro-processors that would fit this requirement. Adafruit has been providing µC boards in a very small form factor that include a high speed CAN interface using DMA. Included on the Feather M4 CAN board is a CAN Transceiver ready to interface with LCC. The only additional hardware needed is the RJ-45 connectors.
2. Programming in an IDE. These Adafruit boards are recognized by the Arduino IDE. What is needed is a simpler library for LCC. Specialized library functions would ease the learning curve for the hobbyist. Providing the "boiler-plate" functions of LCC would get the hobbyist into the part that is unique to him. Start-up, Alias negotiation, and other network messages are necessary, but do not get the hobbyist into the unique actions that are required. Like in the Arduino, these could all be handled by initializing the CAN network with a call to a library function. Included would be the generation of the XML required for CDI.
3. Add-on hardware like the Arduino would provide the additional I/O functions for special devices. Adafruit has a number of specialized boards that plug onto the Adafruit Feather. These provide displays, motor control and input functions.

As users develop special applications, sharing will become possible. I anticipate that there are specialized applications that are only of interest to a few people. These will never be commercially available, but with a simple DIY . they will be possible.

## Onward into the future

LCC continues to grow. I expect to see more manufacturers releasing products. The number of implementations is already large. More will be coming. Enjoy.

# Appendix A References

The following websites are useful references for more detailed information. Use your favorite search engine to search for the name shown in bold type.

1. **NMRA Standards** contain details necessary to implement LCC. The Standards and Technical Notes should be read together. Most of the information necessary to use LCC is in this book. For more details, reference the NMRA Standards.
2. **RR-CirKits** has all the manuals for the Nodes used in this book, plus several clinic charts from conventions. This website is the commercial site for the purchase of the Nodes.
3. **JMRI Signaling** has a lot of information on signaling, both prototypical and model. Many of the terms used in this book are defined or described in JMRI. If you wish to follow a particular prototype railroad's signaling rules, many "Rule Books" are available.
4. **OpenLCB** is the originating group for the LCC design. This website has several sections, including current standard documents, proposed draft standards, and a knowledge base with clinic charts and other documents.
5. **Modular Signal System** includes the concept of Cooperative Detection. This is described in the NMRA Magazine May 2023 and website https://modularsignalsystem.info/index.html
6. Tumble down signals: https://www.logicrailtech.com/apbdemo.htm
7. Full APB Signaling: https://www.lundsten.dk/us_signaling/abs_apb/index.html#:~:text=In%20the%20early%20tumble-down%20scheme%20the%20tumble-down%20is,signals%20in%20the%20face%20of%20the%20oncoming%20train%3A

# Appendix B: LCC Implementation Side Projects

It seems like every project always gets diverted along the way to other needs, leading to sub-projects. Writing a book seems to be no exception. There is always research and testing to perform. Some of these result in a paragraph or two in this book. Others are the unseen backstage support. Here are some of these projects. I have included these to provide starting points in some cases and in other cases, I provide functions that you can use.

## A Simple Power-On Preset

TowerLCC and SignalLCC boards do not store the state for startup. Sometimes very random displays will pop up. If there is a simple button to push that sets a specific state, that is the answer. If, on the other hand, there are several things that need to be set, other options are necessary. Possibly the easiest option is to put in a "Preset" button that sends an event throughout the network to set specific states. This method can also reset everything to a known state if several operators have been using the layout. There should be a delay between the power coming completely on and the preset because the nodes have initializing before they are ready to act on events.

A slightly more elegant and automatic method is a Power-on preset circuit as described here. The idea is to detect when the power is sufficiently on and then send an event to the network. Using a cheap Arduino μC and thee discrete components a "sensor" input can be created for a TowerLCC. The time delay is provided by a RC delay circuit which will be input to the μC as an analog voltage. When the capacitor voltage reaches 67% of the power supply voltage (for a 5 volt supply that is about 3.3 volts) the μC will send a pulse to the TowerLCC. The second resistor ensures that the capacitor is drained when the power is turned off. Starting points for the Capacitor and resistors are r2 = 20K. R1 = 100K and C1 = 100

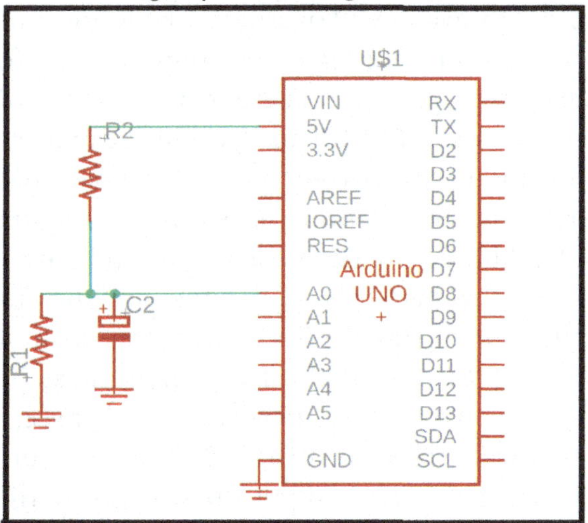


**Figure 14.1** **Add a small μC and a few components to get a Power-On Signal**

μfd. That should be about 1-2 msec. maximum. The highest voltage it will see is about 4 volts, so that would be the longest before a pulse would be sent.

Modifications to this circuit: First, I would use a Nano μC because they are available almost everywhere for little cost. There is no need for much processing power so, use what is cheap and convenient.

Second, if you are low on available inputs, the output could be shared if you add a transistor (preferable a FET) on the output. Then you can use this as a parallel to a button, or as the sensor part of a sampling button,

Third (not a modification really) I have left the program to the implementor. An Arduino program would simply keep looking at the input on A0 to see when it reached a value that correspond to the desired time. Then, a pulse would be sent.

# Kato Turnout Control

In the demonstration layout I wanted to have a simple track construction. I chose Kato Unitrack because it was easy to put together and allowed me to have a two part layout for easy transportation. Unfortunately, Kato did not make it easy to throw their turnouts from LCC. The turnouts are single-coil solenoid style. What that means is that the control for the turnout needs the voltage reversed to change the direction of the turnout, then reversed again to change back to the original direction. This is like a DC motor reversing, except you don't want to keep the voltage on, because the coil will overheat. The Kato turnout control applies the voltage for some short time like 100 milliseconds. To mimic this action, I found a H-bridge breakout board at Adafruit. This board contains a TB6612 H-bridge which can drive two motors or two turnouts. The board has pads for connections as shown in Figure 14.2. Similar boards are available on Ebay.

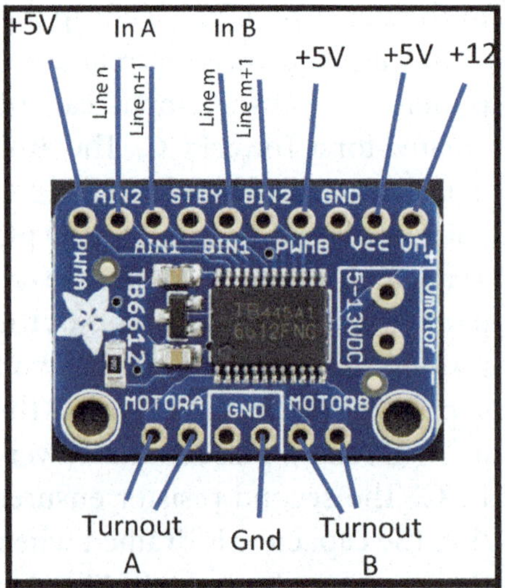


**Figure 14.2 A Basic H-bridge for Kato Turnout Control**

The pads along the top are the input/control signals. To operate the turnout labeled A, one of the two inputs must be turned on for a short period (20 -100 msec.). That will throw the turnout in one direction. Turning the other input on will throw the turnout in the other direction.

Voltage for the logic part of the IC is supplied on the Vcc pad; voltage for the turnout solenoid is supplied on the Vm pad or the Vmotor pad. There are three Gnd pads, which are connected together, so you only have to connect to one, as long as the supplies are connected together outside the board.

The PWM function is disabled by the 5 volts connected to the pads labeled PWMA and PWMB. With the setup in Configuration described in Chapter 6, this is all it takes for direct control of a Kato Turnout.

# Kato Turnout Control Complete

The problem with the previous design is that it uses too many Lines and special features. Requiring two lines to throw the turnout seems wasteful when the total number of nodes is limited. In the Demo layout, the number of nodes is low, so it doesn't matter. In a larger layout the number of nodes could be an issue.

The solution is to use an inexpensive μC to control the turnouts through an H-bridge. The μC I used is called an ItsyBitsy by Adafruit. It is a fast powerful μC and it is inexpensive. Other μCs would be cheaper, but I had two of these. I used the same H-bridge because it is easy to control. Also, this H-bridge has a logic section which operates on the voltage of the μC which is a good way to interface with the μC. The connection to the internal regulator of the ItsyBitsy is the leftmost jumper in Figure 14.3. That jumper brings the logic voltage to the center bus for use to set up the H-bridge.

This prototype is on prototype boards and shows 8 turnout controls. Each μC can control four turnouts with two breakout boards.

Turnout voltage would be input on the terminals on the end of each H-bridge board. 12 volts is a good voltage for these turnouts. The bottom two H-bridges have terminal strips for the turnouts, but simple wires can also be used. Not shown, the inputs from the LCC node will connect to four unconnected lines on the μC.

The outputs are pins A1, A2, A3, A4, 7, 9, 10 and 11. The Arduino sketch reads the input for a turnout to sense a change. When a change occurs, the sketch sets the corresponding output line high to trigger the turnout. After a short time (100 msec is reliable) the output is disabled and the sketch waits for the next change. With this device, the LCC node only has one Line to throw the turnout and does not need to use a pulse output.

# Lurker Projects with Feather CAN

The Adafruit Feather M4 CAN board is almost a perfect μC to connect to the LCC network. As it is designed, it has a rather powerful processor with an integrated CAN network support. Additional devices on the board include the network transceivers, a boost power converter to supply 12 volts for the network and a 2MB Flash memory for persistent storage. The CAN network signals are provided on a three wire terminal strip on the end of the board. The only addition necessary to get connected to LCC is a RJ-45 breakout board. The RJ-45

Breakout board provides the two RJ-45 connectors and a terminal strip to tap into the network. Figure 14.4 shows how to connect the breakout board to the Feather.

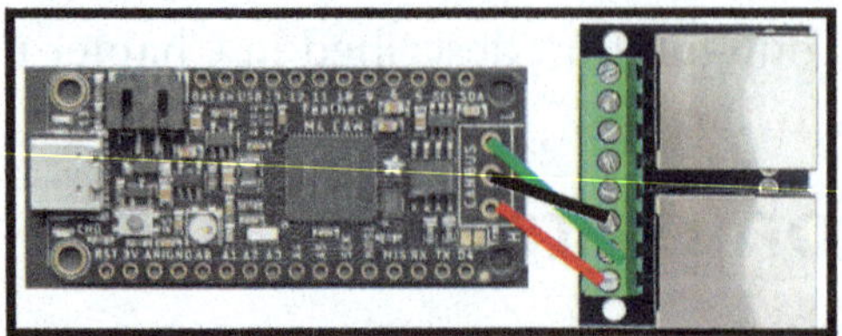

**Figure 14.4 Adafruit Feather Connections to an LCC Network**

One of the advantages of the feather line of boards is that there are many boards that can be plugged onto the pins on the top. These boards, called Feather wings, provide features to interface with LCC. Among these are Multi-digit displays, LCD displays, Motor drivers, GPS receivers and keypads. Adafruit also has a CAN library to use, which partially decodes the messages on the network. The full LCC interface is not yet available, but it will be soon. In the meantime, I have created a Fast Clock Display which I call a Lurker, because it connects to the L:CC network, but it cannot be seen or participate on LCC.

## Fast Clock Display

I started from the Adafruit example for receiving CAN messages. When I connected to the LCC network, I set the CAN bus speed at 125 kps to match LCC. As soon as I started LCC and the Arduino monitor, LCC messages started appearing. Then I continued to select the messages I wanted and displayed them on the serial monitor.

Taking advantage of a seven-segment display offered by Adafruit, I plugged in the display on the connectors on top of the Feather. Then I used the library for the seven-segment display to show the time. This configuration is shown in Figure 14.5. You might notice that the display used here is different than the one shown in Chapter 9. Adafruit offers at least two sizes of displays, and also offers several sizes of LCD displays which can be programmed to show numbers. To change the size, no changes need to be made in the program.

**Figure 14.5 Three wires connect the Lurker Fast Clock Display to LCC**

The Arduino program was a standard structure with a setup section and a loop to keep looking for messages. Several different messages types are used to keep the display in sync with the Fast Clock. The program is available on GitHub at:

*https://github.com/dwzimmerli/LCC4Arduino/tree/main*

This program shows how to use the LCC messages, but not how to interact with LCC.

# Appendix C Rule Book Fragment

*This is a portion of the Belmont Shore Line Rule Book. This fragment is provided as a reference for the signaling implemented in this book. Changes that may occur in the approved BSL Rule Book will not necessarily be incorporated in this fragment.*

## General Description of Signals

### Signal Placement

Signals should be placed where the engineer can see the indication without extreme movement beyond the normal operating location. In most cases, the signal p,lacement should be on the Right-hand side of the track as shown:

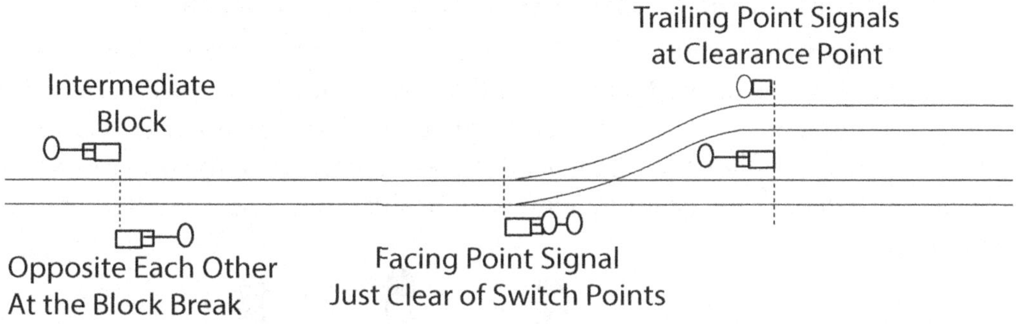


Sidings and signals between tracks may be implemented as dwarf signals. Dwarf signals are signals installed on masts that place the signal head near the top of the rail. Dwarf signals must be installed such that the signal indication is clearly visible.

If the normal placement of a signal mast is not observable when approaching the block controlled by the signal, other locations should be considered. An example would be a switch immediately at the exit of a tunnel. In this case, the signal or a duplicate should be installed at the entrance to the tunnel.

If multiple tracks need signals, the use of bridges or multiple heads on brackets is recommended.

### Signal Implementations

Signals may be implemented with one of three types of Signal Heads. These are the Color Position light, the Searchlight, and the Semaphore. When selecting colors for signals, care must be taken to ensure the color is clearly identified as Red, Yellow or Green. In particular, the Green must be as close to the prototypes Admiralty Green as possible. These signal types are shown below.

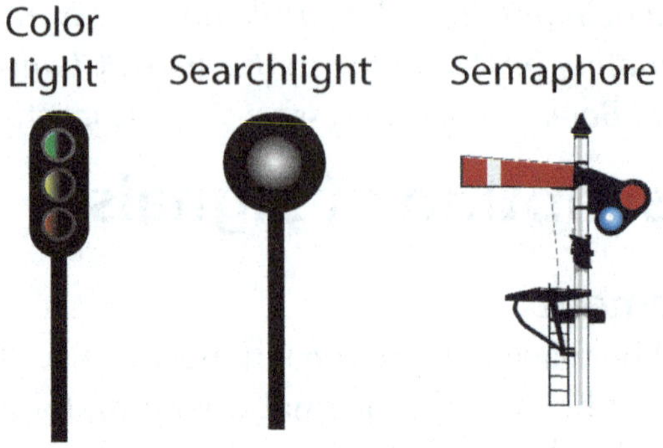


## Signal Installation on Masts, Bridges and Brackets

Signals are normally mounted on masts vertically. If there are two signal heads. the upper head governs the Main line, and the lower head governs the diverging siding or branch line. If more than two signal heads are required, or if a bridge arrangement is installed, the signal heads govern the tracks in the order in which they are installed. Color Position Signals on Bridges may be rotated 90° for added visibility.

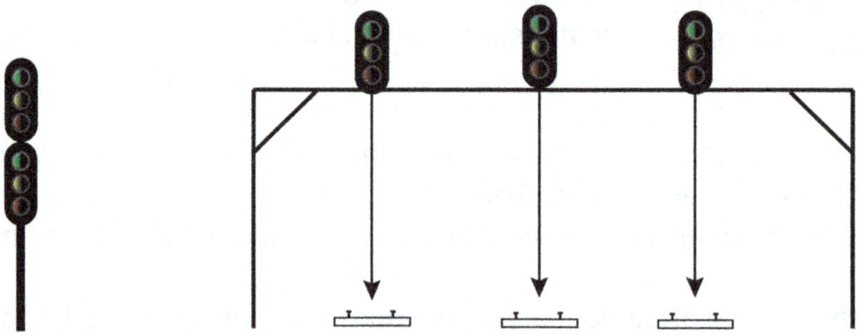

## Signal Rules, Aspects, and Indications

| Aspect | Rule No | Name | Indication |
|---|---|---|---|
| 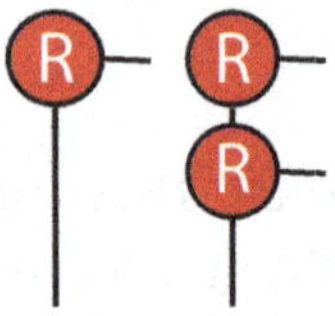 | 54A | Stop | Stop before any part of the train or engine passes the signal. |

| Aspect | Rule No | Name | Indication |
|---|---|---|---|
| 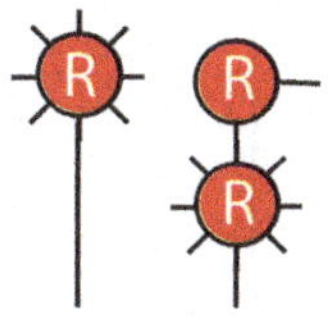  | 54B | Restricted | Stop and wait for less restrictive indication. |
| 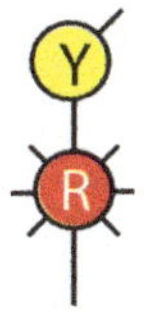  | 54C | Approach Restricted | Proceed prepared to pass next signal at Restricted speed. When proceeding on the Diverging Route, this aspect means that the switch is not set. |
|   | 54D | Approach Diverging | Approach next signal prepared to proceed on diverging route at authorized speed |
|   | 54E | Approach Limited | Proceed prepared to pass next signal not exceeding 60 MPH on diverging route. |
| 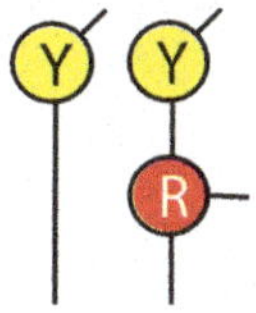  | 54F | Approach | Proceed prepared to stop at next signal; trains exceeding 40 MPH immediately reduce to that speed |
| 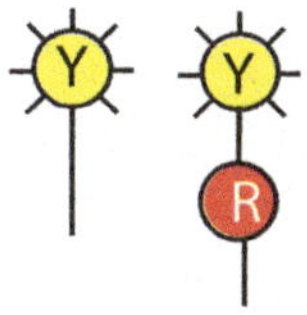  | 54G | Advance Approach | Proceed prepared to pass next signal not exceeding 50 MPH on diverging route |
|   | 54H | Diverging Clear | Proceed on diverging route at authorized speed. |
| 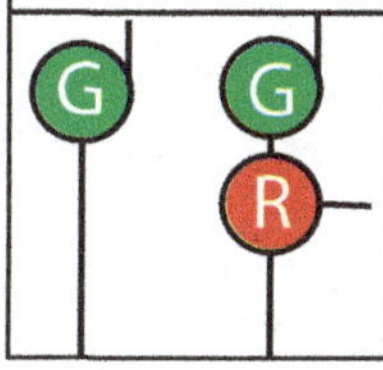  | 54I | Clear | Proceed at Authorized speed on Mainline |

Explanation of Symbols

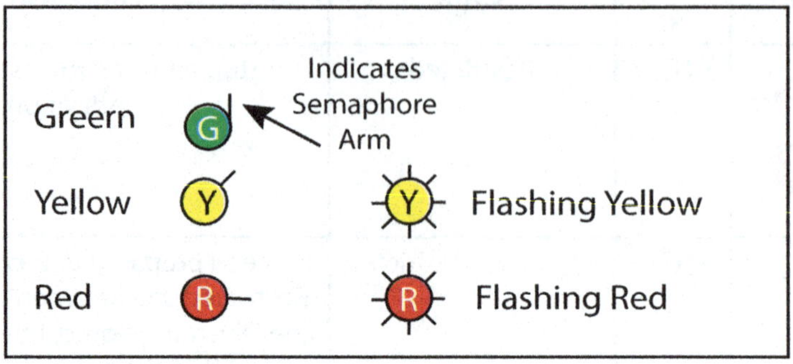

# Index

D



E



F

# About the Author

Dana attended Graduate School at Iowa State University where he obtained a PhD in Computer Science and Electrical Engineering.

Dana retired as a Project Manager from Raytheon in 2011, after working on several high-tech programs from early GPS to the High-flying UAV. After retirement, he became very active in the Belmont Shore Model Railroad club, leading up to his involvement in LCC.

Made in United States
Orlando, FL
05 November 2026

90208348R00142